THE PRACTICAL USE OF THEORY IN FLUID FLOW

Book I

INERTIAL FLOWS

STUART WINSTON CHURCHILL

The Carl V.S. Patterson Professor of Chemical Engineering
The University of Pennsylvania
Philadelphia, Pennsylvania

ETANER PRESS
Education · Technology · the Arts · Natural Science · Ecology · Reports
Thornton, Pennsylvania 19373
1980

THE PRACTICAL USE OF THEORY IN FLUID FLOW
Book I: INERTIAL FLOWS

Library of Congress Catalog Card Number: 79-93367
ISBN: 0-936282-01-0 Volume I; 0-936282-00-2 Series

Printed in the United States of America.
By direct subscription with the publisher.

ETANER Press
P.O. Box
Thornton, PA 19373
USA

Prospectus
for
The Practical Use of Theory in Fluid Flow

The primary objective of the projected series of books with the above general title is to demonstrate the utility of simple theoretical concepts for practical calculations involving the flow of fluids. A collateral objective is to develop the ability of the reader to derive and apply such theoretical results for his own new problems.

The titles of the individual books are:

I. Inertial Flows
II. One-dimensional Laminar Flows
III. The General Equations of Motion
IV. Unconfined Multidimensional Flows
V. Confined Multidimensional Laminar Flows
VI. Confined Turbulent Flows
VII. Flows Through Dispersed Media

All books on fluid flow necessarily reflect a subjective and limited choice of breadth and depth of coverage. The choice in this series is based on the above objectives. The subject matter of the individual books is primarily in order of increasing complexity with some deference to the interrelationship between the topics of Books IV to VII. Since the aim is to develop the skill and confidence of the reader in formulating models, deriving solutions and interpreting theoretical results, these books are reader-oriented rather than topic-oriented.

Although the same objective and style pervade all seven books, they are essentially independent. Book III, or the equivalent, is a prerequisite for Books IV and V. Otherwise the books can be studied selectively or in any order. Developments in the earlier books are utilized in subsequent books but only in the sense of a convenient reference. The division of this treatment of fluid flow into seven separate short books with minimal coupling is intended to provide maximum flexibility and minimal cost in classroom use. This arrangement should also be helpful for self-study.

The books, as indicated by the title of the series, are primarily theoretical rather than empirical. Theoretical concepts are often presumed to be difficult to understand, theoretical derivations to be mathematically demanding and theoretical results to be irrelevant and impractical. These misapprehensions are refuted in this series by countless examples. Theoretical concepts are shown to explain otherwise baffling behavior. Simple derivations from first principles are used to obtain illustrative results of great utility and generality. Theoretical models are demonstrated to be a safer guide to practice than any amount of empiricism.

Approximate models and approximate methods of solution are utilized whenever feasible. Solutions of first-order accuracy are usually sufficient for engineering purposes. More complex models and more involved methods of solution are introduced only when necessary in practical terms. That is, theoretical concepts and mathematical techniques are not introduced for their own sake.

The most significant and obvious omission is a description of numerical methods for the solution of ordinary and partial differential equations. Such methods are being used increasingly to solve models for fluid flow. They are generally more practical than classical analytical methods. The author has worked on the development of numerical methods throughout his career and is partial to them. However, these methods are highly specialized and in a rapid state of development. The best method for one problem may be poor for another. Those used today may be out of date tomorrow. Therefore, these methods themselves are not discussed or illustrated. Of course, results so obtained are included wherever appropriate.

A significant fraction of the material in this series of books is unique, i.e., it has not appeared in any other book. Examples are the experimental and computed values for detonation velocities in Book I and the methods of dedimensionalization and correlation which are utilized in Books II, IV, V, VI and VII.

The notes on which these monographs are based were developed over a number of years for the first half of a one-semester course in momentum and heat transfer for seniors and first-year graduate students in chemical engineering.* These classes have, however, generally included a significant fraction of practicing engineers and students in contiguous fields such as aeronautical engineering, bioengineering, mechanical engineering and chemistry. A primary consideration has, therefore, been the wide diversity of preparations in thermodynamics, fluid mechanics and heat transfer, ranging from zero through superficial acquaintance to real understanding from student to student and topic to topic.

Students appear to learn most readily by first examining very simple phenomena and models and then considering the effect of added complexities. Hence that procedure has been followed. Equations and solutions for one-dimensional, limiting cases are first derived. Then terms and dimensions are gradually added. Reduction of unsteady-state, three-dimensional partial differential equations to simple cases does not appear to achieve the same rapidity or degree of understanding.

Is this material of graduate level? The answer will vary with the respondent. Many graduate courses are exercises in pride rather than education. They are sophisticated, highly mathematical and narrow in scope. They are based on the premise that the students already understand the basic material to which they were exposed in their undergraduate curriculum. On the contrary, I have found that most first-year graduate students do not have a working knowledge of the elementary concepts of fluid flow. Hence the degree of review in this volume is not only acceptable but essential if comprehension rather than mere exposure is the goal. The material is equally appropriate for most seniors since they generally have the same preparation in fluid mechanics as first-year graduate students.

The Problem Sets constitute an essential part of the material. Detailed solution of a reasonable number and distribution of the problems is necessary for comprehension. Furthermore, some topical developments are deferred completely to the Problem Sets. A manual of complete solutions of all of the problems in each book will be published shortly after the book itself.

---

*S. W. Churchill, "Theories, Correlations and Uncertainties for Waves Gradients and Fluxes—a Course in Momentum and Energy Transfer," *Chem. Eng. Educ.* **3**, 178, 179, 181, 182 (Fall, 1969).

ACKNOWLEDGEMENTS

This book grew out of a course begun with Donald L. Katz and James G. Knudsen and later taught jointly with James O. Wilkes, J. Louis York, Dudley A. Saville and Joe D. Goddard at the University of Michigan. They all contributed to the material herein.

The contributions of my doctoral students, Martin Gluckstein and Roy Gealer, as well as those of Morton Moyle, are evident.

Special acknowledgement is also appropriate to another former student, Noel de Nevers, whose book on fluid mechanics was a continual reminder that these topics could be treated in words as well as equations, and without pomposity.

Generations of students at the University of Michigan and the University of Pennsylvania have provided constructive criticisms and suggestions.

Marcia Goldberg typed the preliminary manuscript with skill, forbearance and good humor.

Permissions

To

Morton P. Moyle, who first stimulated my interest in many of these topics and whose career was terminated so untimely.

Preface

This book examines fluid flows for the limiting case of high velocities such that viscous effects can be neglected relative to inertial effects. This inertial regime of flow is characterized by high Reynolds numbers and/or high Mach numbers. Although viscous effects remain significant near surfaces, the portion of the flow which is affected may be negligible, permitting the further, associated simplification of one-dimensionality for a filament of fluid between two adjacent streamlines, even though the streamlines are two or three-dimensional. The assumptions of inviscid flow and one-dimensionality greatly simplify the mass and force-momentum balances and in many cases permit the use of algebraic rather than differential models. The linear differential equations which characterize purely viscous flow are thus replaced by nonlinear algebraic equations. Even superficial comparison of the material in Books I and II of this series confirms that inertial flows are easier to predict and correlate than those in which viscous effects predominate.

Multidimensional, inviscid and *incompressible* flow, commonly called *ideal* fluid flow, forms the principal content of many theoretical and classical books on fluid mechanics, such as those by Milne-Thomson [1], Streeter [2] and Lamb [3]. *Compressible,* inviscid flows are the principal concern of books on theoretical gas dynamics such as those by Cambel and Jennings [4] and Emmons [5]. The restricted but important applications of these theoretical results to chemical and physical processing have received remarkably little attention. Some of the handbooks and unit operations books even infer a purely empirical basis for results which can be derived directly from first principles.

An objective of this book is to bring together those particular aspects of one-dimensional, inertial flow which are applicable for the design and operation of processing plants. Both compressible and incompressible flows are considered. The scope and value of the results which are obtained under these severe restrictions are truly remarkable. However, since education is a further and more important objective, emphasis is on the formulation and solution of models rather than on the results in themselves. References are provided for two-dimensional treatments and other more advanced aspects.

The particular topics covered can be summarized as follows. One-dimensional mass and force-momentum balances are first shown to lead to the form of the well-known expressions for Venturi tubes, orifices, weirs and nozzles. An expression for the acoustic velocity is next derived. The acoustic velocity is shown to provide a constraint on flow through a nozzle. The results for nozzles are demonstrated to be applicable for performance calculations for rockets and jets. The maximum velocity in a pipe is also shown to be sonic. Expressions for sudden contractions and expansions and for the permanent loss due to an orifice are shown to follow directly from one-dimensional balances. The concept of a step in velocity, pressure and temperature is used to derive expressions for shock, detonation and gravity waves. Finally cavitation, incipient vaporization and aerodynamic heating are examined. All of these results are obtained without empiricism.

As indicated by the above summary, the principal value of the results herein for inertial flow is for the prediction and explanation of limitations on flow and processing which are not evident from the better known empirical correlations and solutions for purely viscous flow. Indeed, this book might justifiably be entitled *Inertial Limitations on Flow.* For example, sonic velocity at the throat of a nozzle or at the exit of a pipe provides an upper limit on the possible rate of flow of a gas. Similarly a standing wave limits the attainable velocity in open-channel flow of a liquid. Cavitation, vaporization and aerodynamic heating are also primarily of interest because of the restriction they apply to flows.

References

1. L. H. Milne-Thompson, *Theoretical Hydrodynamics,* 2nd Ed., Macmillan, New York (1950).
2. V. L. Streeter, *Fluid Dynamics,*McGraw Hill, New York (1948).
3. H. Lamb, *Hydrodynamics,* 1st American Edition, Dover, New York (1945).
4. A. B. Cambel and B. J. Jennings, *Gas Dynamics,* McGraw Hill, New York (1958).
5. H. W. Emmons, Editor, *Fundamentals of Gas Dynamics,* Princeton University Press, Princeton, New Jersey (1958).

A note on units

The original material in the text of this book follows the *International System of Units (Le Système Internationale d'Unités)*, hereafter abbreviated SI, for all dimensional quantities. However, most thermodynamic and fluid dynamic data in the literature are still to be found in English or cgs units. Furthermore, most engineering practice will be carried out for some time in English or cgs units. Hence it is necessary for engineers to be proficient in conversion from one set of units to another. In recognition of this state of affairs the problems at the end of each chapter generally reflect the form in which the data were found. Also, plots of data have of course been reproduced in their original units. A short table of the pertinent units, constants and conversion factors for the SI system is provided in **Appendix B.**

If the base units of the SI system are used, no conversion factors are required between force, energy, inertia, etc. However, the use of special-name units such as the newton (for force), the joule (for energy), the pascal (for pressure) and the watt (for power) introduces redundancy and hence the necessity of conversion factors. With one exception, the units of the SI were chosen so that all of these conversion factors, such as newton-meters per second to watts, have a numerical value of unity. That exception is the mole, which is defined for a gram equivalent rather than a kilogram equivalent of a chemical species or compound. Hence care must be taken in expressions and calculations which involve both mole and mass units to avoid an error of a factor of 10^3. One way to avoid this difficulty is to use the *molar mass* (kg/mol), herein symbolized by m, rather than the *molecular weight* in all such expressions and calculations. For example, the molar mass of carbon-12 is 0.012 kg/mol.

Contents

Chapter 1

Reversible expansions and compressions

Momentum balance

Newton's second law of motion* equates the time rate of change of momentum with the sum of the applied forces. It is applicable to fluid motion if a segment of fluid is treated as a rigid body. In general, momentum is a second-order tensor and separate expressions must be written for each of the three orthogonal directions. However, the restriction of attention to processes that can be treated as one-dimensional permits a scalar application of Newton's law.

Consider a filament of non-viscous fluid moving steadily along a streamline as shown in Fig. 1.1. The tube represents the boundaries of this segment of fluid as it changes direction and velocity. The applied forces will here be restricted to gravity and pressure. Newton's second law can be written over a differential length of this "rigid" filament as follows:

$$-dP - g\rho dz = \rho u du \tag{1.1}$$

where

P = absolute pressure, Pa or N/m^2 or $kg/m{\cdot}s^2$
g = acceleration (or specific force) due to gravity, m/s^2 or N/kg
ρ = density, kg/m^3
z = vertical component of distance travelled, m
u = velocity, m/s

**LAW II*
The change of motion is proportional to the motive force impressed; and is made in the direction of the right line in which that force is impressed.

If any force generates a motion, a double force will generate double the motion, a triple force triple the motion, whether that force be impressed altogether and at once, or gradually and successively. And this motion (being always directed the same way with the generating force), if the body moved before, is added to or subtracted from the former motion, according as they directly conspire with or are directly contrary to each other; or obliquely joined, when they are oblique, so as to produce a new motion compounded from the determination of both. [1]

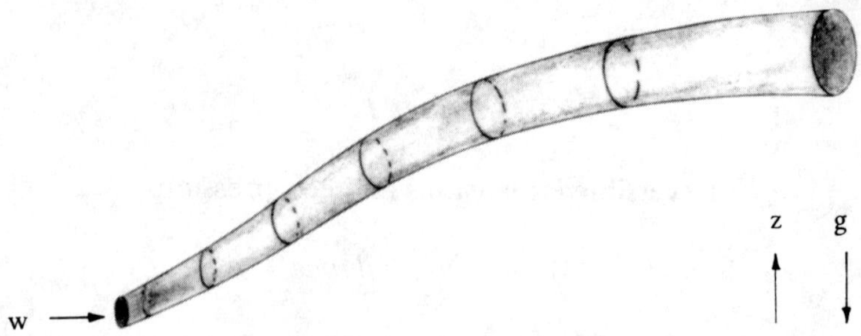

Fig. 1.1. Moving filament of fluid

Equation (1.1) can be derived by reduction of the general equations of motion of a fluid. (See, for example, Bird et al. [2]). It is known as the differential form of the *Bernoulli equation.* However, Equation (1.1) sometimes is called the *Euler equation* and the name Bernoulli reserved for the special case of incompressibility. The development following Equation (1.1) is sometimes called *filament theory.*

Integrating formally between any two locations 1 and 2 yields

$$\frac{u_2^2 - u_1^2}{2} + g(z_2 - z_1) + \int_{P_1}^{P_2} \frac{dP}{\rho} = 0 \tag{1.2}$$

Mass balance

Since the total mass rate of flow through the imaginary tube in Fig. 1.1 does not change with length

$$dw = d(u\rho A) = 0 \tag{1.3}$$

where

w = mass rate of flow through area A, kg/s
A = cross-sectional area, m^2.

It follows that

$$\frac{du}{u} + \frac{d\rho}{\rho} + \frac{dA}{A} = 0 \tag{1.4}$$

Integration of Equation (1.3) gives

$$w = uA\rho = u_1 A_1 \rho_1 = u_2 A_2 \rho_2 \tag{1.5}$$

Combined mass and momentum balances

Elimination of u_2 between Equations (1.2) and (1.5) and rearrangement gives

$$u_1^2 = \frac{2\left[g(z_1 - z_2) + \int_{P_2}^{P_1} \frac{dP}{\rho}\right]}{\left(\frac{\rho_1 A_1}{\rho_2 A_2}\right)^2 - 1} \tag{1.6}$$

or

$$w^2 = \frac{2(\rho_2 A_2)^2\left[g(z_1 - z_2) + \int_{P_2}^{P_1} \frac{dP}{\rho}\right]}{1 - \left(\frac{\rho_2 A_2}{\rho_1 A_1}\right)^2} \tag{1.7}$$

Equations (1.6) and (1.7) provide the basis for the correlation of data for Venturi tubes, orifices and nozzles. In order to apply these equations a relationship is needed between P and ρ. Several important cases are examined in the next section.

Evaluation of the integral of expansion in steady flow

The density is in general a function of temperature and composition as well as of pressure so that some restrictions or specifications are necessary to evaluate the integral of expansion $\int_{P_2}^{P_1} \frac{dP}{\rho}$ even when the thermodynamic behavior of the fluid is known. Attention here will be limited to several special cases. One or the other of these cases is generally an acceptable approximation in applications.

Incompressible fluids. If ρ is constant

$$\int_{P_2}^{P_1} \frac{dP}{\rho} = \frac{P_1 - P_2}{\rho} \tag{1.8}$$

and

$$u_1^2 = \frac{2\left(g(z_1 - z_2) + [(P_1 - P_2)/\rho]\right)}{(A_1/A_2)^2 - 1} \tag{1.9}$$

The assumption of incompressibility is a reasonable one for gases as well as liquids if the change in density is not too great and a mean density, ρ_m, is chosen such that

$$\int_{P_2}^{P_1} \frac{dP}{\rho} \equiv \frac{P_1 - P_2}{\rho_m} \tag{1.10}$$

If the variation of ρ is monotonic, as is often the case, ρ_m must lie between ρ_1 and ρ_2.

Isothermal flow of an ideal, non-reacting gas. For an ideal gas

$$\rho = Pm/RT \tag{1.11}$$

where

m = molar mass, kg/mol

T = absolute temperature, K

R = universal gas law constant = 8.3143 J/mol · K
or $kg \cdot m^2/s^2 \cdot mol \cdot K$ or $m \cdot N/mol \cdot K$ or $Pa \cdot m^3/mol \cdot K$.

In the absence of a chemical reaction, m is a constant and

$$\int_{P_2}^{P_1} \frac{dP}{\rho} = \int_{P_2}^{P_1} \frac{RT}{Pm} dP = \frac{RT}{m} \ln \left\{ \frac{P_1}{P_2} \right\} \tag{1.12}$$

or

$$\int_{P_2}^{P_1} \frac{dP}{\rho} = \frac{RT}{m} \frac{\ln \{P_1/P_2\}}{P_1 - P_2} (P_1 - P_2) = \frac{RT(P_1 - P_2)}{mP_{lm}} = \frac{P_1 - P_2}{\rho_{lm}} \tag{1.13}$$

where

$$P_{lm} = \text{logarithmic-mean } P \equiv (P_2 - P_1)/\ln\{P_2/P_1\} \tag{1.14}$$

$$\rho_{lm} = \text{logarithmic-mean } \rho \equiv (\rho_2 - \rho_1)/\ln\{\rho_2/\rho_1\} \tag{1.15}$$

The heat input required to maintain constant temperature during an expansion or compression is calculated as follows. For steady flow the first law of thermodynamics can be written in the form

$$q - w = \Delta H + \Delta\left(\frac{u^2}{2}\right) + g\,\Delta z + \ldots \tag{1.16}$$

where

q = heat entering system per unit of mass entering, J/kg or m^2/s^2

w = work done by system per unit of mass entering, J/kg or m^2/s^2

H = specific enthalpy, J/kg or m^2/s^2

Δ = finite difference between State 2 (leaving) and State 1 (entering)

$+ \ldots$ = the effect of body forces other than gravity.

ΔH_T, the change in specific enthalpy at constant temperature, can be shown to be zero for an ideal gas as follows. The enthalpy can be expressed as a function of any two intensive variables such as pressure and temperature. Then since

$$dH = \left(\frac{\partial H}{\partial P}\right)_T dP + \left(\frac{\partial H}{\partial T}\right)_P dT \tag{1.17}$$

$$dH_T = \left(\frac{\partial H}{\partial P}\right)_T dP \tag{1.18}$$

Also since

$$dH = TdS + dP/\rho \tag{1.19}$$

where

$$S = \text{specific entropy, J/kg} \cdot \text{K or m}^2/\text{s}^2\,\text{K}$$

it follows that

$$\left(\frac{\partial H}{\partial P}\right)_T = T\left(\frac{\partial S}{\partial P}\right)_T + \frac{1}{\rho} \tag{1.20}$$

The Maxwell relationships indicate that

$$\left(\frac{\partial S}{\partial P}\right)_T = -\left(\frac{\partial(1/\rho)}{\partial T}\right)_P \tag{1.21}$$

For an ideal gas, from Equation (1.12)

$$\left(\frac{\partial(1/\rho)}{\partial T}\right)_P = \frac{R}{P_m} = \frac{1}{T\rho} \tag{1.22}$$

Substituting from Equation (1.22) in (1.21) and then from (1.21) in (1.20) indicates that

$$\left(\frac{\partial H}{\partial P}\right)_T = 0 \tag{1.23}$$

It follows from Equation (1.18) that

$$\Delta H_T = 0 \tag{1.24}$$

This result is of course restricted to an ideal gas. For isothermal free expansion (no work) of an ideal gas, Equation (1.16) then reduces to

$$q_T = \Delta\left(\frac{u^2}{2}\right) + g\,\Delta z \tag{1.25}$$

From Equations (1.25), (1.3) and (1.12)

$$q_T = \int_{P_2}^{P_1}\left(\frac{dP}{\rho}\right) = \frac{RT}{m}\ \ln\left\{\frac{P_1}{P_2}\right\} \tag{1.26}$$

Further, from Equation (1.19)

$$q_T = \int_{S_1}^{S_2} TdS = T(S_2 - S_1) \tag{1.27}$$

The heat input required to maintain constant temperature during a free expansion or compression is thus seen to be equal to the sum of the increases in kinetic and potential energy, also to $(RT/m)\ \ln\{P_1/P_2\}$ and also to $T\,\Delta S$.

Adiabatic, reversible flow of an ideal, nonreacting gas. For free, adiabatic flow q and w are zero. For reversible flow $\Delta S = 0$. Considering entropy to be a function of pressure and density only requires

$$dS = (\partial S/\partial P)_\rho\, dP + (\partial S/\partial \rho)_P\, d\rho = 0 \tag{1.28}$$

It follows that

$$(\partial S/\partial T)_\rho\,(\partial T/\partial P)_\rho\, dP = -\,(\partial S/\partial T)_P\,(\partial T/\partial \rho)_P\, d\rho \tag{1.29}$$

By definition the specific heat capacity at constant pressure, in J/kg · K or

m^2/s^2 K, is

$$c_P \equiv (\partial H/\partial T)_P \tag{1.30}$$

From Equation (1.19)

$$c_P = T(\partial S/\partial T)_P \tag{1.31}$$

Similarly the heat capacity at constant volume (or density) is

$$c_V \equiv (\partial U/\partial T)_\rho = T(\partial S/\partial T)_\rho \tag{1.32}$$

where $U = H - P/\rho$ = internal energy per unit mass, J/kg or m^2/s^2 or $Pa \cdot m^3/kg$

From Equation (1.12)

$$(\partial T/\partial P)_\rho = m/R\rho \tag{1.33}$$

and

$$(\partial T/\partial \rho)_P = - Pm/R\rho^2 \tag{1.34}$$

Substituting from Equations (1.31) - (1.34) in Equation (1.29):

$$c_V \, mdP/TR\rho = c_P \, mPd\rho/TR\rho^2 \tag{1.35}$$

Simplifying and rearranging

$$c_V \, dP/P = c_P \, d\rho/\rho \tag{1.35a}$$

Calling $c_P/c_V = \gamma$, the heat capacity ratio, and integrating for constant γ gives

$$\ln\{P\} = \gamma \ln\{\rho\} + B \tag{1.36}$$

where B is a constant.

Hence

$$P/\rho^\gamma = P_1/\rho_1^\gamma = P_2/\rho_2^\gamma = \text{a constant} \tag{1.37}$$

It follows from (1.11) that

$$\frac{\rho}{\rho_1} = \left(\frac{T}{T_1}\right)^{\frac{1}{\gamma - 1}} = \left(\frac{P}{P_1}\right)^{\frac{1}{\gamma}} \tag{1.38}$$

Then

$$\int_{P_2}^{P_1} \frac{dP}{\rho} = \int_{P_2}^{P_1} \left(\frac{P_1}{P}\right)^{1/\gamma} \frac{dP}{\rho_1} = \frac{\gamma P_1{}^{1/\gamma}}{\rho_1 (\gamma - 1)} \left(P_1{}^{\frac{\gamma - 1}{\gamma}} - P_2{}^{\frac{\gamma - 1}{\gamma}} \right) \tag{1.39}$$

$$= \frac{\gamma}{\gamma - 1} \frac{P_1}{\rho_1} \left[1 - \left(\frac{P_2}{P_1}\right)^{\frac{\gamma - 1}{\gamma}} \right] \tag{1.39a}$$

$$= \frac{\gamma}{\gamma - 1} \left(\frac{P_1}{\rho_1} - \frac{P_2}{\rho_2} \right) \tag{1.39b}$$

From Equations (1.12) and (1.39b) it follows that

$$\int_{P_2}^{P_1} \frac{dP}{\rho} = \left(\frac{\gamma}{\gamma - 1}\right) \frac{R}{m} (T_1 - T_2) \tag{1.40}$$

For an ideal gas

$$(c_P - c_V)m = R \tag{1.41}$$

The derivation of Equation (1.41) can be found in all texts on thermodynamics. Equation (1.41) permits simplification of Equation (1.40) to

$$\int_{P_2}^{P_1} \frac{dP}{\rho} = c_P (T_1 - T_2) \tag{1.40a}$$

For an adiabatic process without work, Equation (1.16) reduces to

$$\Delta\left(\frac{u^2}{2}\right) + g\Delta z = -\Delta H \tag{1.42}$$

Combining Equations (1.3), (1.40a) and (1.42) results in

$$\Delta\left(\frac{u^2}{2}\right) + g\Delta z = -\Delta H = c_P(T_1 - T_2) \tag{1.43}$$

Thus the decrease in enthalpy, sensible heat ($c_P\Delta T$) and temperature for free, isentropic expansion of an ideal, nonreacting gas corresponds to the increase in kinetic and potential energy.

From Equation (1.43) it follows that for horizontal flow

$$T = T_1 + \frac{u_1^2 - u^2}{2c_P} \tag{1.44}$$

If the flow is stopped completely, the resulting, limiting temperature is called the *stagnation temperature* and is defined as

$$T_s = T_1 + \frac{u_1^2}{2c_P} \tag{1.45}$$

It is apparent from Equation (1.45) that the stagnation temperature is invariant in adiabatic flow even though the velocity, pressure and flowing temperature are changing. Hence this is a useful quantity for the calculation or interpretation of adiabatic flows. T_s is sometimes called the *reservoir temperature* corresponding to T_1 and u_1. Conversely Equation (1.45) gives the temperature T_1 attained by a gas accelerated from rest at T_s to u_1.

As indicated by Problem 1.1, Equations (1.42) - (1.45) are applicable for irreversible as well as reversible flow.

For adiabatic, *reversible*, horizontal flow of an ideal, nonreacting gas it follows from Equation (1.38) that the stagnation pressure and stagnation density are also invariant.

Other expansions and compressions. Nonisothermal or nonadiabatic expansions and compressions are often approximated by the empirical expression

$$P/\rho^k = \text{constant} \tag{1.46}$$

It may be noted that $k = 1$ corresponds to an isothermal expansion and $k = \gamma$ corresponds to Equation (1.37) for an adiabatic expansion. Most actual expansions are intermediate between isothermal and adiabatic, i.e., $1 < k < \gamma$.

Heat capacity ratio. Approximate representative values of the heat capacity ratio γ for gases are given in Table 1.1. More precise values of γ or of c_P and c_V as functions of temperature and pressure may be obtained from tabulations, graphs or empirical equations for these or other related thermodynamic functions. The heat capacity ratio varies moderately with temperature as illustrated for air and methane in Table 1.2 and somewhat less with pressure except near the critical pressure. However, these variations may be magnified since $\gamma - 1$ appears in many of the prior and subsequent expressions herein for expansions and compressions.

Table 1.1

Approximate, representative values of γ at 1 atm and 298 K

Gas	γ
C_2H_6, C_2H_4, C_2H_2	~ 1.2
CH_4, H_2O*, H_2S, NH_3, Cl_2, CO_2	~ 1.3
Air, H_2, N_2, O_2, CO, NO and HCl	~ 1.4
Monatomics	~ 1.67

* *at saturation pressure*

Table 1.2

Variation of γ of air and methane with temperature

T, °C	0	100	200	300	400	500	600	700	800	900	1000	2000	3000
Air	1.410	1.397	1.398	1.378	1.367	1.356	1.347	1.338	1.331	1.325	1.319	1.243	1.283
CH_4	1.318	1.274	1.228	1.196	1.174	1.158	1.145	1.136	1.129	1.122	1.117		

Applications involving small changes in density

The above expressions for reversible expansions and compressions are applicable to nozzles and orifices as long as the velocity does not exceed that of sound. Such low-velocity applications are considered in Chapter 2.

Problems

1.1. Are Equations (1.40a), (1.42), (1.43), (1.44) and (1.45) restricted to reversible flow? Explain.

1.2. Air is expanded isentropically from 3.0 MPa and 293 K to atmospheric pressure. Calculate the final temperature, and the change in enthalpy, momentum and kinetic energy per kg of air.

1.3. Air is expanded isothermally from 3.0 MPa and 293 K to 1.0 KPa. Calculate the heat input and the change in entropy, kinetic energy and momentum per kg of air.

1.4. Derive an expression for the theoretical maximum attainable velocity in an isentropic expansion. Calculate this velocity for hydrogen initially at 288 K and atmospheric pressure.

1.5. Calculate the exit diameter of a nozzle for the steady discharge of 22 kg/s of air to the atmosphere from a vessel at 1.0 MPa and 298 K. Isentropic flow may be assumed.

1.6. A hot gas stream with $\gamma = 1.35$ and $m = 26$ is expanded through a turbine from 1033 K and 0.5 MPa to the atmosphere. Calculate the work which can be extracted if the expansion is isentropic. What is the final temperature? (Hint: assume $\Delta(u^2/2)$ is negligible).

1.7. Air is compressed isothermally from 60°F and 1 atm to 7 atm. How much heat must be extracted per pound assuming $\Delta(u^2/2)$ is negligible. How much work must be done?

1.8. Derive Equation (1.37) starting from $S = f\{P,T\}$ instead of $S = f\{P,\rho\}$.

1.9. Derive Equation (1.37) starting from $H = f\{P,\rho\}$ instead of $S = f\{P,\rho\}$.

1.10. Size a motor for a fan exhausting 5,000 ft^3/min of 70°F air through a short duct, 12 inches in diameter. Assume an efficiency of 60 per cent for the fan and 85 per cent for the motor.

1.11. The American Falls of the Niagara River are approximately 50 m high. Calculate the temperature increase or decrease incurred by the water in passing over the falls. The effects of evaporation and air-entrainment may be neglected. Note any assumptions.
What is the maximum power in kilowatts which could be extracted by diverting 0.45 Gg/s through a standpipe and a turbine? Note any assumptions.

References

1. Isaac Newton, *Principia*, 2nd Ed., S. Pepys Press, London (1713); English translation by Andrew Motte (1729), revised by Florian Cajori, Vol. I, University of California Press, Berkeley, California (1966), p. 13.
2. R. B. Bird, W. E. Stewart and E. N. Lightfoot, *Transport Phenomena*, Wiley, New York (1960), pp. 211–213.

Chapter 2

Expansion at low velocity

As subsequently demonstrated in Chapter 3, acoustic velocity provides an upper limit to the direct applicability of the expressions developed in Chapter 1 for reversible expansions. However, liquids rarely attain sonic velocities in practical applications and many metering devices such as Venturi nozzles and sharp-edged orifices ordinarily operate below the acoustic limit even with gases. Such subsonic applications are considered in this chapter.

Approximations

In most metering devices the pressure and density changes are generally so small, even for gases, that a mean value of the density can be assumed without serious error.

For horizontal flow through a channel with a round cross section, Equation (1.9) can be simplified to

$$u_1^2 \cong 2(-\Delta P)/\rho_m\,[(D_1/D_2)^4 - 1] \tag{2.1}$$

where

$$D = \text{diameter, m}$$

For a large ratio of D_1/D_2, the further approximation

$$u_1^2 \cong 2(D_2/D_1)^4\,(-\Delta P)/\rho_m \tag{2.2}$$

is sometimes made.

Venturi tubes

A Venturi* tube is a smooth converging-diverging section which is inserted in a pipe to determine the volumetric rate of flow. The pressure drop is usually measured with a manometer with one tap upstream and the other at the throat as indicated schematically in Figure 2.1.

Equation (2.1) holds reasonably well for Venturi tubes but a correction factor is customarily added as a refinement giving

---

**According to Dryden et al., p. 477, this device was first utilized for the measurement of flow by Herschell [2], who named it after Venturi [3], who in 1799 noted the marked decrease in pressure in the constricted part of a tube. However, this phenomenon is also said to have been known to the Romans.*

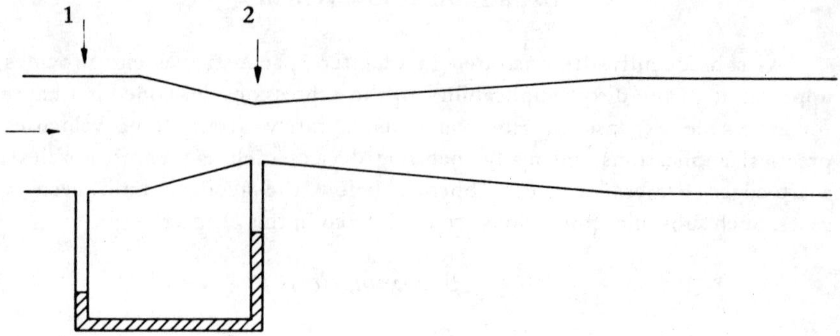

Fig. 2.1. Venturi meter and manometer

$$u_1^2 = 2C_v^2(-\Delta P)/\rho_m\,[(D_1/D_2)^4 - 1] \tag{2.3}$$

with the dimensionless correction factor C_v correlated as a function of the Reynolds number in the pipe = $Re_1 = D_1 u_1 \rho_m/\mu$ and Venturi shape. Values of C_v typically range from 0.9 to 0.99. The deviation of C_v from unity is primarily due to failure of the stream to follow the channel rather than to viscous effects. Avoidance of separation by the use of a smooth tapering inlet, a rounded throat and a long tapering outlet are the key to proper design. Brown et al. [4] assert that the total interior entrance angle is usually 25 degrees (0.436 rad) and the total interior exit angle is 7 degrees (0.1222 rad), and that C_v is 0.98 for $D_1 u_1 \rho_m/\mu > 100{,}000$.

ASME long-radius, converging nozzle

The ASME long-radius flow-nozzle is a standard device used to meter liquids. It consists of a converging section whose length is D_2 and a straight section whose length is 0.6 D_2. The discharge coefficient C_n, defined the same as C_v in Equation (2.3), was found by Benedict [5] to be a function of only $D_2 u_2 \rho/\mu$ for all D_2/D_1, as indicated in Figure 2.2.

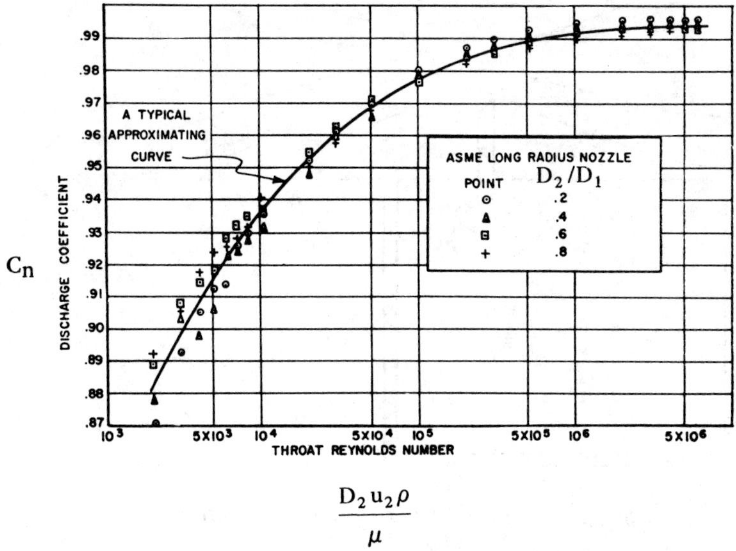

Fig. 2.2. Discharge coefficient for ASME long-radius flow-nozzle [5]

Sharp-edged orifices

Orifices usually consist of a plate with a central hole, sharp-edged upstream and rounded-outward downstream, as illustrated in Figure 2.3. The fluid stream necks down to a diameter less than the sharp-edged diameter of the orifice somewhat downstream and then reexpands. The fluid outside this jet stream is comparatively stagnant but circulates slowly. The point of minimum diameter is known at the *vena contracta*. The orifice plate should be at least 50 pipe diameters downstream and 10 pipe diameters upstream from any fitting or valve to avoid dissymmetry or disturbances in the flow.

To determine the flow from a pressure measurement a coefficient is ordinarily added to Equation (2.1), just as for Venturi nozzles. In terms of the points of pressure measurement, then

$$u_1^2 = 2C_2^2(P_1 - P_2)/\rho_m[(D_1/D_2)^4 - 1)] \qquad (2.4)$$

The coefficient C_2 approaches unity. However D_2, the diameter of the jet of

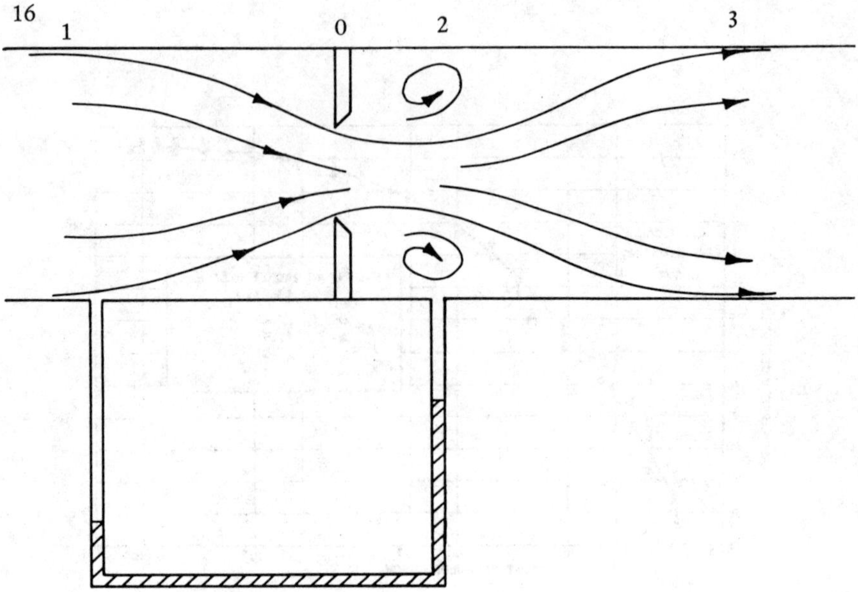

Fig. 2.3. Sharp-edged orifice and manometer

fluid at point-of-measurement-2, is not easily measured. Hence Equation (2.4) is rewritten in terms of the known orifice diameter as follows:

$$u_1^2 = 2C_o^2[P_1 - P_2]/\rho_m[(D_1/D_o)^4 - 1] \qquad (2.5)$$

From comparison of Equations (2.4) and (2.5)

$$C_o^2 = C_2^2[(D_1/D_o)^4 - 1]/[(D_1/D_2)^4 - 1] \qquad (2.6)$$

$$\cong C_2^2(D_2/D_o)^4 \cong (D_2/D_o)^4 \qquad (2.7)$$

Lamb [6] credits Kirchoff in 1869 with the derivation of $C_o = \pi/(\pi + 2) = 0.6110$ for $D_o/D_1 \rightarrow 0$ with pressure-tap-2 located at the *vena contracta.* For well-designed systems in which pressure-tap-2 is located near the *vena contracta* the experimental values of C_o approach an asymptotic value of 0.61 for large $D_o u_o \rho_m/\mu$ confirming the reversibility of flow through an orifice. For laminar flow through a sharp-edged orifice, Roscoe [7] derived the expression

$$u_o = D_o(P_1 - P_2)/6\pi\mu \qquad (2.8)$$

which is equivalent to

$$C_o = (Re_o/12\pi)^{1/2} \qquad (2.9)$$

where

$$Re_o = D_o u_o \rho / \mu = \text{Reynolds number for orifice}$$
$$\mu = \text{dynamic viscosity, Pa} \cdot \text{s or kg/m} \cdot \text{s}$$

This result is an asymptotic solution for $D_o/D_1 \to 0$. Some parametric dependence on D_o/D_1 would be expected as D_o/D_1 increases. This is observed in Figure 2.4, which is a plot by Tuve and Sprenkle [8] of data for "corner connections" – pressure taps at the two surfaces of the orifice plate. The deviations from 0.61 in turbulent flow are primarily due to location of the point of measurement of the upstream pressure too near the orifice, and displacement of the point of downstream pressure measurement from the *vena contracta*, rather than to viscous losses. Churchill and Usagi [9] fitted an empirical equation to the data for $D_1/D_o = 2.5$ using 0.61 and Equation (2.9) as asymptotes.

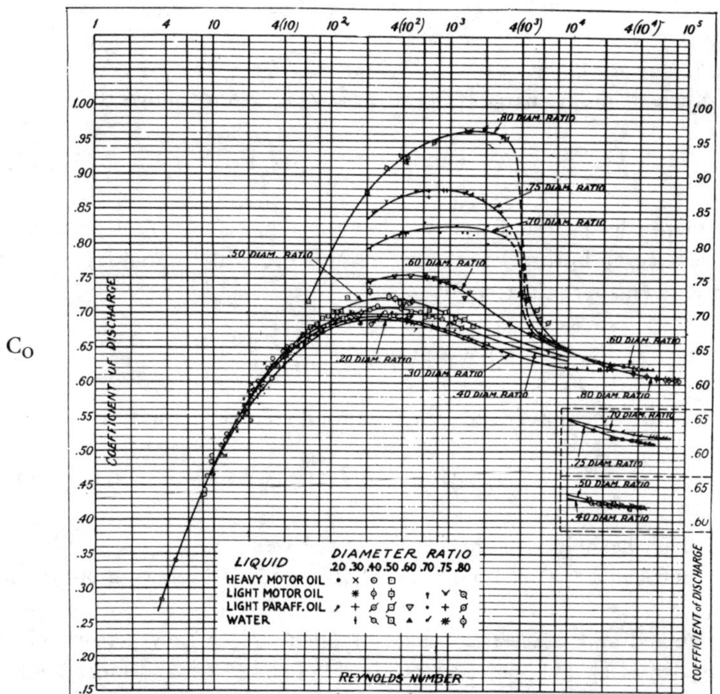

Fig. 2.4. Discharge coefficients for orifice plates with corner connections [8]

A slightly different plot would be obtained for different points of pressure measurement. Experimental values of the orifice coefficient for a variety of designs and conditions have been reviewed by Iverson [10]. Smoothed, standard values have been tabulated by the ASME [11].

Equation (2.5) and Fig. 2.4 provide a reasonable approximation for other geometries if the corresponding value of A_1/A_o is substituted for $(D_1/D_o)^2$.

Free-surface expansions

Approximate expressions for the flow of liquids through slots and over crests as in Figs. 2.5 - 2.7 corresponding to those for orifices can readily be derived as follows.

For a sharp-edged sluice gate, as in Fig. 2.5, Equation (1.9) is applicable with $P_1 = P_2$. In this case $A_1/A_2 = z_1/z_2$ and $z_2 = C_d d$ where d is the open height of the gate and C_d is the discharge coefficient for the gate. Then

$$u_d^2 = 2C_d^2 g z_1/[1 + (dC_d/z_1)] \tag{2.10}$$

where u_d is the mean velocity under the gate.

If the downstream discharge is flooded as in Fig. 2.6 it follows that

$$u_d^2 = 2C_d^2 g(z_1 - z_2)/[1 - (dC_d/z_1)^2] \tag{2.11}$$

For a sharp-edged crested weir as shown in Fig. 2.7, wherein the overflow has a free surface on both sides, the velocity of each differential layer of fluid at the crest is given approximately by

$$u_z = C_d(2gz)^{1/2} \tag{2.12}$$

where here z is the height over the weir. The total volumetric flow rate is

$$v = s\int_0^d C_d(2gz)^{1/2}\,dz = (2^{3/2}/3)C_d\, g^{1/2}\, d^{3/2}\, s \tag{2.13}$$

where here d is the height of the pool over the weir and s is the width of the weir. The coefficient C_d in Equations (2.10) – (2.13) equals approximately 0.61 as $b/d \rightarrow \infty$.

More detailed and exact treatments of open-surface flow across weirs are given by Dryden et al. [1], p. 451 f, Rouse [12], p. 370 f, and most books on hydraulics or open-channel flow.

Isentropic expansion in a nozzle

Consider horizontal, isentropic flow of an ideal gas in a convergent-divergent nozzle as subsequently illustrated in Fig. 3.3. The inlet area will be

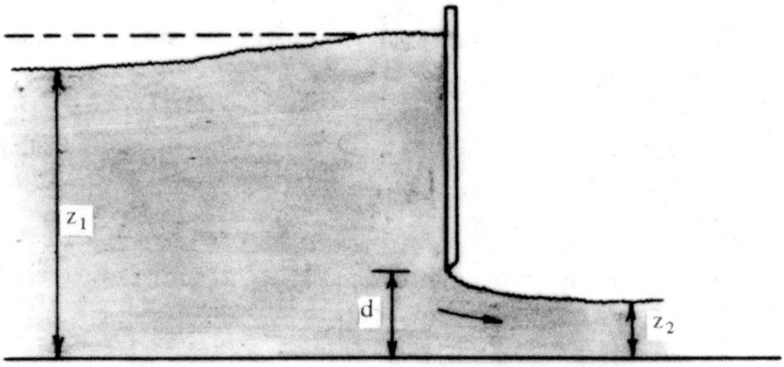

Fig. 2.5. Unflooded sluice gate

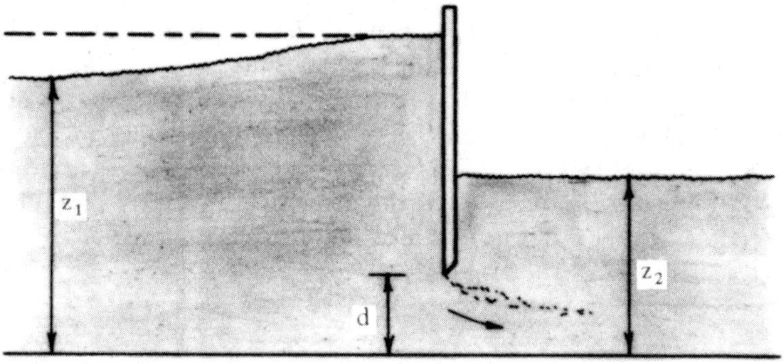

Fig. 2.6. Flooded sluice gate

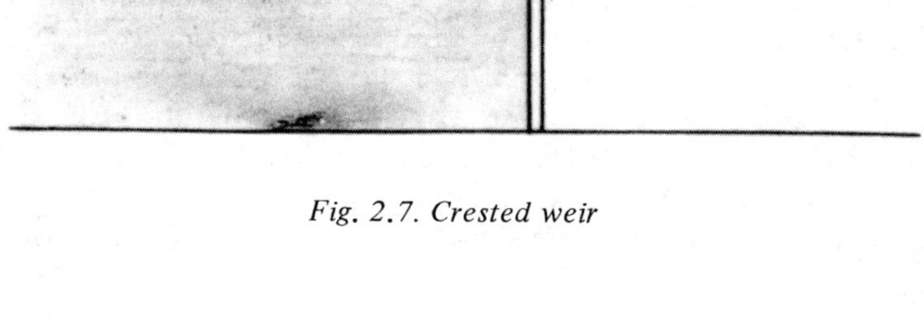

Fig. 2.7. Crested weir

assumed to be sufficiently large so that u_1 can be neglected in Equation (1.2). Combining this reduced form of Equation (1.3) with (1.39a) and replacing u_2 and P_2 with any general u and P gives

$$u^2 = \left(\frac{2\gamma}{\gamma-1}\right)\frac{P_1}{\rho_1}\left(1-\left(\frac{P}{P_1}\right)^{\frac{\gamma-1}{\gamma}}\right) \tag{2.14}$$

Equation (2.14) provides the velocity at any point in the nozzle in terms of the pressure. Substituting $w/\rho A$ for u and $\rho_1 (P/P_1)^{1/\gamma}$ for ρ gives

$$w^2 = A^2\left(\frac{2\gamma}{\gamma-1}\right)P_1\rho_1\left(1-\left(\frac{P}{P_1}\right)^{\frac{\gamma-1}{\gamma}}\right)\left(\frac{P}{P_1}\right)^{\frac{2}{\gamma}} \tag{2.15}$$

The mass rate is thus fixed by $P_1/\rho_1, \gamma$ and any pair of values of P and A such as at the exit. The dimensionless rate $w/A_2(P_1\rho_1)^{1/2}$ is a function only of γ and the pressure ratio P_2/P_1, where here the subscript 2 indicates any location such as the throat or exit. Equation (2.15) also gives the relationship between P and A for a fixed flow rate. The density and temperature are related to pressure through Equations (1.37) and (1.12) and the velocity through (2.14). If the area is known as a function of length these quantities can all be calculated as functions of length. Since the area first decreases then increases, the pressure behaves similarly, reaching a minimum value at the throat.

Problems

2.1. What is the diameter of the *vena contracta* as a fraction of D_o?

2.2. What error arises from the use of Equation (2.2) instead of (2.1) if $D_1/D_2 = 2.0$?

2.3. Estimate the mass flow rate of water through a 100-mm pipe if the manometer reads 3.9 mm Hg
   a) for a Venturi tube with a throat diameter of 50 mm;
   b) for a sharp-edged orifice with $D_o/D_1 = 0.4$, using Fig. 2.4;
   c) for case (b), using the empirical equation given by Churchill and Usagi [8].

2.4. If an orifice is misinstalled with the sharp edge downstream will the indicated flow be too high or too low? Estimate the fractional change.

2.5. Water is maintained at a depth of 2 m behind a sluice gate. Calculate the velocity through an opening 0.6 m high created by lifting the gate. A discharge coefficient of 0.61 may be assumed.

2.6. Rederive Equation (2.5) for a compressible fluid.

2.7. Roscoe [7] gives the expression $-\Delta P = 32\mu u_o/\pi b$ for laminar flow through a slot of width b. Derive an expression for C_O.

2.8. Oil (ρ = 0.68 Mg/m^3) is flowing at 0.111 *l*/s through a Venturi tube with a throat of 10.16 mm in a 52.5 mm pipe. Assuming a coefficient of 0.98, what will be reading in inches measured by a Hg-oil manometer?

2.9. Saturated steam at 1 MPa is to be introduced through taps along a transfer line containing oil vapor and powdered catalyst to prevent clogging. What size orifice should be installed in the taps to produce a flow rate of 67 mg/s?

2.10. Compare the time required to drain water and gasoline through a hole in the bottom of a tank.

2.11. Derive Equation (2.11). Show that Equation (2.11) reduces to (2.10) for no flooding.

2.12. Derive an expression for a square orifice equivalent to Equation (2.5).

2.13. Derive an expression for a long horizontal slot and compare with Equations (2.10) and (2.11).

2.14. Explain Equation (2.12).

2.15. If a river is flowing at 900 m^3/s how high will it crest with respect to a dam 200 m wide?

2.16. How high should the weir height be to achieve a depth of 15 mm over a bubble-cap tray? The reflux rate is 60 *l*/s and the width of the weir is 3 m. The effect of the gas may be neglected.

2.17. How high should a 5-m-long sluice gate be lifted to achieve a flow rate of 30 m^3/s with a pool height of 20 m

a) if the downstream channel is unflooded?

b) if the downstream depth is 15 m?

References

1. H. L. Dryden, F. P. Murnaghan and H. Bateman, *Hydrodynamics,* Dover, New York (1956).
2. C. Herschell, "The Venturi Water Meter," *Trans. ASCE* **17**, 228-258 (1887); 133-141 (1888).
3. G. B. Venturi, *Recherches experimentales,* Paris (1799).
4. G. G. Brown and Associates, *Unit Operations,* Wiley, New York (1950), p. 157.
5. R. P. Benedict, "Most Probable Discharge Coefficients for ASME Flow Nozzles," *J. Basic Engng.* **88D**, 734-744 (1966).
6. H. Lamb, *Hydrodynamics,* 1st American Edition, Dover, New York (1945), pp. 98-99.
7. R. Roscoe, "The Flow of Viscous Fluids Around Plane Obstacles," *Proc. Physical Soc.* 4, 338-351 (1951).
8. G. E. Tuve and R. E. Sprenkle, "Orifice Discharge Coefficients for Viscous Liquids," *Instruments* 6, 201-206 (1933).
9. S. W. Churchill and R. Usagi, "A Standardized Procedure for the Pro-

duction of Correlations in the Form of a Common Empirical Equation," *Ind. Eng. Chem., Fundam.* **13**, 39-44 (1974).

10. H. W. Iverson, "Orifice Coefficients for Reynolds Numbers from 4 to 50,000," *Trans. ASME* **78**, 359-364 (1956).
11. *ASME Power Test Code 19*, Supplement on Instruments and Apparatus, Part 5, Measurement of Quantity of Materials, Chapter 4, *Flow Measurement by Means of Thin Plate Orifices, Flow Nozzles and Venturi Tubes* (1959).
12. H. Rouse, *Engineering Hydraulics*, Wiley, New York (1959).

Chapter 3

Maximum reversible rates of flow for a gas

The flow of gas through orifices, nozzles and pipes is limited by the attainment of sonic velocity at the *vena contracta*, throat and exit, respectively. Such "choked" flows occur widely in industrial practice but are not always anticipated or recognized. This behavior is the most frequent cause of underdesign in piping systems. Limiting, reversible flows of gas are examined in this chapter. The application of these relationships to rocket and jet nozzles is discussed in Chapter 4. The limiting flow of gas in pipes, for which the irreversibilities cannot be neglected, is considered in Chapter 5.

Velocity of a weak pressure wave

Consider a compression wave moving at steady velocity a through a stagnant, compressible material (gas, liquid or solid) at an initial pressure P_1 and initial density ρ_1 producing a velocity du, a pressure $P_1 + dP$ and a density $\rho_1 + d\rho$ behind the wave as indicated in Fig. 3.1.

The equations of motion are simplified if the frame of reference for velocity is changed from the surroundings (or unperturbed material) to the wave. This is accomplished simply by subtracting the wave velocity from the velocities in Fig. 3.1. Since the wave velocity is greater than the particle velocities the new velocities, as designated by primes, are positive toward the left and are drawn in this sense in Fig. 3.2. P and ρ are omitted from Fig. 3.2 since they are unaffected by the change in the frame of reference.

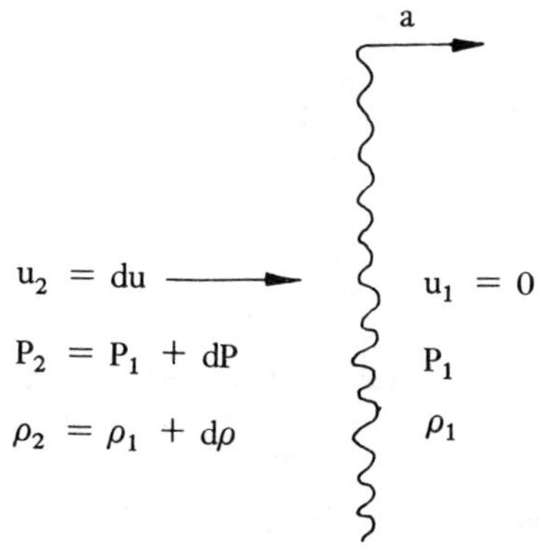

Fig. 3.1. Weak compression wave

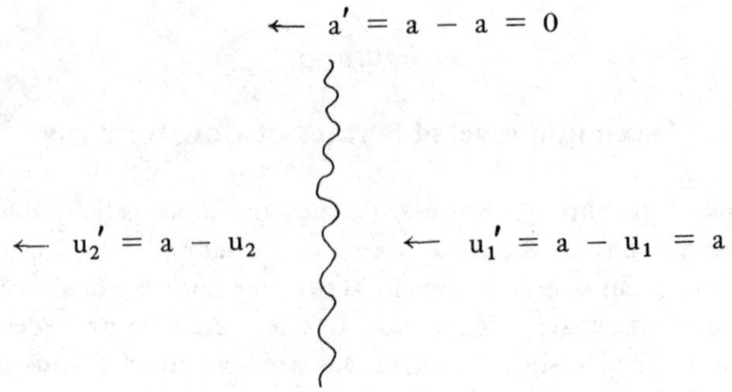

Fig. 3.2. Particle velocities relative to a stationary compression wave

The material balance for the process described in Fig. 3.2 takes the form

$$u_1'\rho_1 = u_2'\rho_2 \tag{3.1}$$

or in terms of the original velocities and densities

$$a\rho_1 = (a - du)(\rho_1 + d\rho) \tag{3.2}$$

Similarly a force and momentum balance gives

$$P_1 + (u_1')^2\rho_1 = P_2 + (u_2')^2\rho_2 \tag{3.3}$$

or

$$P_1 + a^2\rho_1 = P_1 + dP + (a - du)^2\,(\rho_1 + d\rho) \tag{3.4}$$

Eliminating du between Equations (3.2) and (3.4) and simplifying:

$$a^2\rho_1\,d\rho = \rho_1\,dP + d\rho dP \tag{3.5}$$

Dropping the second-order differential yields

$$a^2 = dP/d\rho \tag{3.6}$$

Equation (3.6) can also be derived directly from Equations (1.1) and (1.4) [see Problem 3.31] or from the general equations for the conservation of mass and momentum.

Equation (3.6) indicates that for an isothermal wave in an ideal gas

$$a_T^2 = (\partial P/\partial \rho)_T = RT/m = P/\rho \tag{3.7}$$

and for an isentropic wave in an ideal gas

$$a_S^2 = (\partial P/\partial \rho)_S \tag{3.8}$$

Substituting for P from Equation (1.37) gives

$$a_S^2 = \partial[P_1(\rho/\rho_1)^\gamma]/\partial\rho = \gamma P_1 \rho^{\gamma-1}/\rho_1^\gamma = \gamma P/\rho = \gamma RT/m \tag{3.9}$$

Equations (3.7) and (3.9) can be generalized as

$$a^2 = kRT/m = kP/\rho \tag{3.10}$$

where $k = 1.0$ for an isothermal wave and γ for an isentropic wave.

The measured acoustic velocity in gases is in close agreement with Equation (3.9).* The velocity of an acoustic wave in any material can in principle be calculated from Equation (3.8). a_T and a_S differ negligibly for liquids and solids and the experimental data are generally expressed in terms of the bulk modulus,

$$E \equiv a_T^2 \rho \tag{3.11}$$

in kg/m·s^2, rather than directly in terms of a_S or a_T.

Equation (3.8) was derived for and is applicable only for small changes in pressure. The velocity generated by large pressure differences exceeds the acoustic velocity. Expressions for strong waves are derived in *Chapter 6*. Acoustic velocity is actually the lowest velocity at which a pressure wave can propagate.

The Mach number

The dimensionless ratio of the velocity to the acoustic velocity, u/a, is

---

**According to Westfall [1], Newton derived Equation (3.7), then both manipulated the experimental data and deliberately misinterpreted the equation to force agreement. Laplace recognized that the process was necessarily adiabatic and arrived at Equation (3.9).*

Gorring [2] asserts that for nonideal gases

$$a_S^2 = Z^2 \gamma RT/m \tag{3.9a}$$

where Z is the compressibility factor. Squire [3] has proposed an approximate expression for use near the critical point. His prediction is in good agreement with experimental data for ethylene. Whiting and Ackerberg [4] have recently tested other methods of prediction for nonideal gaseous mixtures.

called the *Mach number** and given the symbol M. When using M it is important to identify the velocity, including the frame of reference, and the conditions (e.g., the particular temperature and molar mass or pressure and density) for which *a* is calculated.

For horizontal flow Equation (1.1) reduces to

$$dP/\rho + udu = 0 \tag{3.12}$$

Substituting for dP from Equation (3.6) gives

$$a^2\, d\rho/\rho + udu = 0 \tag{3.13}$$

Combining Equations (3.13) and (1.5) to eliminate $d\rho/\rho$:

$$a^2 \left(\frac{du}{u} + \frac{dA}{A} \right) = udu \tag{3.14}$$

or

$$\frac{du}{u} \left[1 - \left(\frac{u}{a} \right)^2 \right] = - \frac{dA}{A} \tag{3.14a}$$

Introducing the Mach number:

$$\frac{du}{u}\,[1 - M^2] = - \frac{dA}{A} \tag{3.15}$$

From Equation (3.15) it follows that

for $M < 1$: u decreases as A increases and vice versa,

for $M > 1$: u increases as A increases and vice versa.

Equation (3.14a) cannot be integrated in general since M depends on T which depends on u. However, a particular example of the relationship between u and A is provided in the following section.

Limiting flow in an isentropic nozzle

The same conditions will be postulated as in the derivation of Equation (2.11): horizontal flow of an ideal gas with constant γ and with a large enough initial area so that u_1^2 can be neglected. Rearranging Equation (2.15):

$$A^2 = \frac{w^2(\gamma - 1)}{2\gamma P_1 \rho_1} \left(\left(\frac{P}{P_1} \right)^{\frac{2}{\gamma}} - \left(\frac{P}{P_1} \right)^{\frac{\gamma+1}{\gamma}} \right)^{-1} \tag{3.16}$$

**This dimensionless ratio is named in honor of the Austrian physicist, Ernst Mach, who demonstrated the role of the acoustic velocity in aerodynamics [5].*

Equation (3.16) gives the variation of P with A, or A with P. For a given w, P_1 and ρ_1, A first decreases from infinity at $P = P_1$ and then increases without limit as $P \to 0$. Hence some minimum (throat) area, A_t, must exist at some critical pressure-ratio P_c/P_1. A_t and P_c/P_1 can be found by setting

$$d(A^2)/d(P/P_1) = 0 \tag{3.17}$$

Carrying out this operation on Equation (3.16) gives

$$\frac{2}{\gamma}\left(\frac{P_c}{P_1}\right)^{\frac{2-\gamma}{\gamma}} - \frac{\gamma+1}{\gamma}\left(\frac{P_c}{P_1}\right)^{\frac{1}{\gamma}} = 0 \tag{3.18}$$

or

$$\frac{P_c}{P_1} = \left(\frac{2}{\gamma+1}\right)^{\frac{\gamma}{\gamma-1}} \tag{3.18a}$$

Substitution of P_c from Equation (3.18a) into Equation (3.16) yields

$$A_t^2 = \frac{w^2}{\gamma P_1 \rho_1}\left(\frac{\gamma+1}{2}\right)^{\frac{\gamma+1}{\gamma-1}} \tag{3.19}$$

Also, from Equations (3.18a) and (1.38) the critical temperature and density ratios are

$$\frac{T_c}{T_1} = \left(\frac{\rho_c}{\rho_1}\right)^{\gamma-1} = \frac{2}{\gamma+1} \tag{3.20}$$

Further, the critical velocity at the throat is

$$u_c^2 = \left(\frac{w}{A_t \rho_c}\right)^2 = \frac{\gamma P_1 \rho_1}{\rho_1{}^2}\left(\frac{2}{\gamma+1}\right)^{\frac{\gamma+1}{\gamma-1} - \frac{2}{\gamma-1}} = \frac{2\gamma P_1}{(\gamma+1)\rho_1} \tag{3.21}$$

In terms of conditions at the throat

$$u_c^2 = \left(\frac{2\gamma}{\gamma+1}\right) P_c \left(\frac{\gamma+1}{2}\right)^{\frac{\gamma}{\gamma-1}} \left(\frac{1}{\rho_c}\right)\left(\frac{2}{\gamma+1}\right)^{\frac{1}{\gamma-1}} = \frac{\gamma P_c}{\rho_c} \tag{3.22}$$

The critical velocity at the throat is seen to be acoustic. This is the maximum velocity that can be attained in a converging nozzle or in the converging section of a converging-diverging nozzle. The maximum velocity, as indicated by Equation (3.21), depends only on the upstream conditions. A physical explanation for the occurrence of acoustic velocity and the independence from downstream conditions is that a small wave cannot propagate upstream against a gas velocity greater than sonic and hence the effect of downstream changes cannot be communicated upstream.

The maximum mass flow rate obtained by rearrangement of Equation (3.19) is

$$w_c^2 = \gamma P_1 \rho_1 A_t^2 \left(\frac{2}{\gamma+1}\right)^{\frac{\gamma+1}{\gamma-1}} \tag{3.23}$$

Converging nozzle

In a converging nozzle the limiting flow indicated by Equation (3.23) is attained whenever P_c is greater than the external pressure P_∞, thus

$$\frac{P_1}{P_\infty} > \frac{P_1}{P_c} = \left(\frac{\gamma+1}{2}\right)^{\frac{\gamma}{\gamma-1}} \tag{3.24}$$

This ratio is independent of the conditions of flow other than γ and is essentially a constant for a whole group of fluids. For air and other diatomic gases with $\gamma \cong 1.4$

$$\frac{P_c}{P_1} \cong \left(\frac{2}{1.4+1}\right)^{\frac{1.4}{1.4-1}} = 0.528$$

and the required downstream pressure is approximately one-half the upstream pressure.

If $P_c > P_\infty$: the flow rate is given by Equation (3.23). The gas expands down to P_c at the throat and then irreversibly to P_∞ outside the nozzle.

If $P_c < P_\infty$: the flow rate is given by Equation (2.15) with P_∞ substituted for P and A_t for A. The gas expands monotonically down to P_∞ at the throat.

Equation (3.16) has two real, positive roots, i.e., two values of P for each A. The greater root corresponds to the expansion in the converging section and also to subsonic recompression in the diverging section. The lesser root corresponds to supersonic expansion in the diverging section. The choice between these two cases is decided by the external pressure as illustrated in Fig. 3.3. Let P_3 and P_5 be the greater and lesser roots of Equation (3.16) for A_e and w_c (as given by Equation (3.23)). The rate of flow w falls in one of the following categories depending on the external pressure P_∞:

If $P_\infty = P_2 > P_3$ the rate of flow is given by Equation (3.16) for A_e and P_∞, and will be less than w_c. The pressure will vary with A as given by Equation (3.16), hence as

$$\left(\frac{A}{A_e}\right)^2 = \left[\left(\frac{P_\infty}{P_1}\right)^{\frac{2}{\gamma}} - \left(\frac{P_\infty}{P_1}\right)^{\frac{\gamma+1}{\gamma}}\right]\left[\left(\frac{P}{P_1}\right)^{\frac{2}{\gamma}} - \left(\frac{P}{P_1}\right)^{\frac{\gamma+1}{\gamma}}\right]^{-1} \tag{3.25}$$

The velocity will be subsonic at the throat.

2) If $P_\infty \leqslant P_3$ the rate of flow is w_c, as given by Equation (3.23), and the gas will expand to P_c in the converging section according to

$$\left(\frac{A}{A_t}\right)^2 = \frac{\gamma-1}{2}\left(\frac{2}{\gamma+1}\right)^{\frac{\gamma+1}{\gamma-1}}\left[\left(\frac{P}{P_1}\right)^{\frac{2}{\gamma}} - \left(\frac{P}{P_1}\right)^{\frac{\gamma+1}{\gamma}}\right]^{-1} \tag{3.26}$$

which results from elimination of w between (3.16) and (3.23). The expansion then falls in one of the following four subcategories depending again on the external pressure P_∞:

a) If $P_\infty = P_3$ the gas will recompress subsonically in the diverging section in accordance with the greater root of Equation (3.16). This behavior is called *ideal subsonic expansion.*

b) If $P_\infty = P_5$ the gas will expand supersonically in the diverging section in accordance with the lesser root of Equation (3.16). This behavior is called *ideal supersonic expansion.*

c) If $P_\infty = P_6 < P_5$ the gas will expand supersonically in the diverging section down to P_5 at the exit in accordance with the lesser root of Equation (3.16) just as for P_5 and then irreversibly to P_6 outside the nozzle. This behavior, which is called *under-expansion*, is economical in rocket nozzles owing to the reduction in the length and exit diameter of the nozzle, and hence in the fabrication cost, drag, etc.

d) If $P_\infty = P_4$ and $P_5 < P_4 < P_3$ the gas will expand supersonically in the diverging section down to some pressure P_7 in accordance with the lesser root of Equation (3.16) as for P_5, then shock recompress to P_8 and then expand

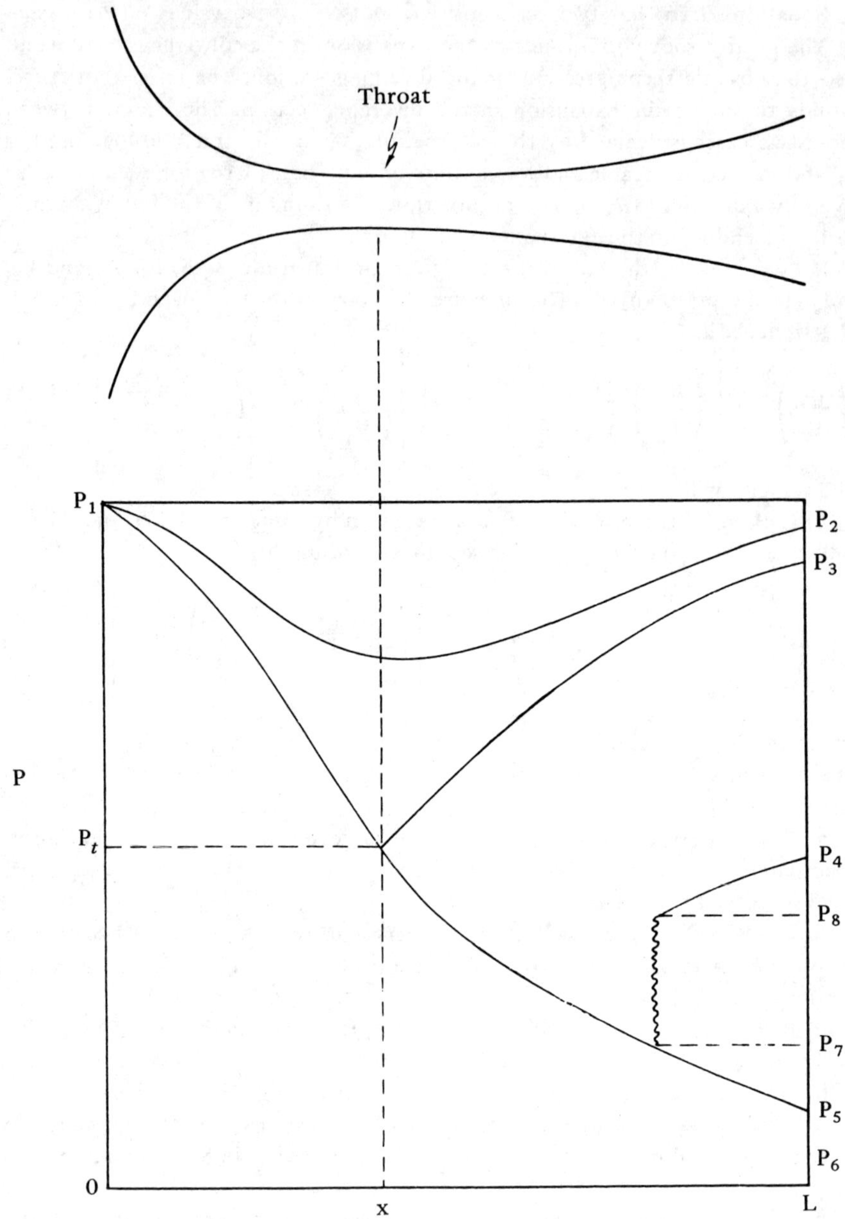

Fig. 3.3. Pressure profiles in a nozzle

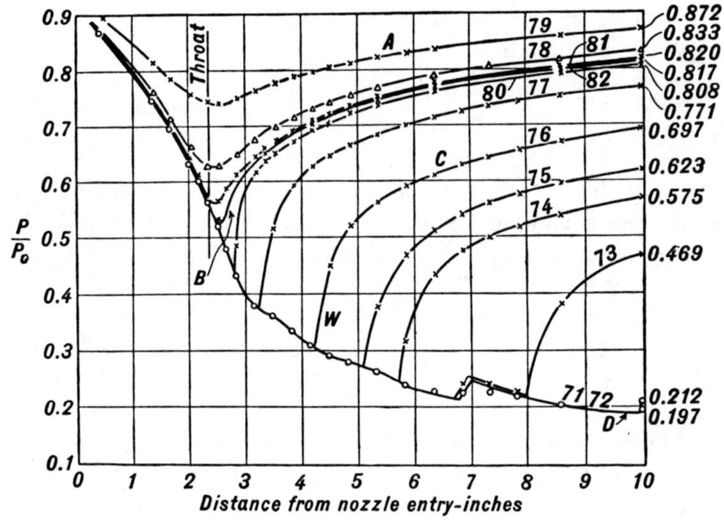

Fig. 3.4. Experimental data of Binney and Woods [6] for expansion of steam through a converging-diverging nozzle

isentropically and subsonically to P_4 in accordance with the greater root of Equation (3.26). This behavior is called *over-expansion* and is very undesirable since noise and vibration are produced by the shock wave. P_7 and P_8 can be calculated by simultaneous solution of the equations in *Chapter 6* for shock compression from P_7 to P_8 together with Equation (3.16) for the segments $P_c - P_7$ and $P_8 - P_4$.

Real nozzles

The experimental data of Binney and Woods [6] for the expansion of steam through a converging-diverging nozzle are shown in Fig. 3.4. Curves B and D represent approximately the upper and lower design curves for critical flow. Some nonideality is evident in curves B and above in that the minimum pressure occurs slightly beyond the throat. The steep portion of the intermediate curves such as labelled W on C indicates a shock wave. The break in curves 71, 72 and 73 at 7 inches was caused by condensation of the steam and is not a shock wave, as indicated by the subsequent shock wave in curve 73. The curves to the left of the throat can be interpreted as the behavior for a converging-only nozzle for external pressures corresponding to the pressure at the throat.

The shape of the nozzle, i.e., the relationship between the cross-sectional area and length is not determined by the above equations. The change in area must be sufficiently gradual to prevent separation. The converging section may be quite steep but the throat should be well-rounded and the total internal angle of the diverging section should not exceed 7 degrees. Deviations from the idealized treatment herein due to viscosity, two-dimensional effects and separation are discussed in detail by Emmons [7].

Orifices

Grace and Lapple [8] found experimentally that the limiting rate of flow through orifices was somewhat less than the value given by Equation (3.23). Accordingly, they introduced an empirical coefficient resulting in

$$w_c^2 = C_{oc}^2 \gamma P_1 \rho_1 A_t^2 \left(\frac{2}{\gamma+1}\right)^{\frac{\gamma+1}{\gamma-1}} \tag{3.27}$$

For sharp-edged orifices they obtained values of 0.82-0.86 for C_{oc} as $A_t/A_1 \to 0$. Reversing the orifices to put the sharp edge downstream gave $C_{oc} \cong 0.91$. Such orifices with the sharp edge downstream are often used with Equation (3.23) to measure rates of flow under the name "critical flow prover". The results of Grace and Lapple and of Godt [9] indicate that calibration may be necessary.

Problems

3.1. Calculate the velocity of a small pressure wave in 288 K air assuming, (a) isentropic compression and (b) isothermal reversible compression. Compare these values with experimental values of the acoustic velocity.

3.2. Calculate the velocity of sound at 288 K and atmospheric pressure in air, hydrogen and helium.

3.3. Calculate the velocity of sound in liquid water using values of P and ρ in the steam tables. Look up the bulk modulus, calculate the acoustic velocity and compare with the prior value.

3.4. Estimate the rate of leakage of gas through a 3-mm hole in a pipeline containing natural gas (90% CH_4 and 10% C_2H_6) at 2.75 MPa and 300 K.

3.5. In a pressure test the pressure in a 0.14-m^3 vessel fell from 0.34 MPa to 0.17 MPa in 24 hours. What rate of leakage would be expected at 0.135 MPa?

3.6. Air is observed to leak at a rate of 1.26 g/s from a large vessel maintained at 10 MPa.

a) Estimate the rate at which hydrogen would leak from the same vessel.
b) Estimate the rate at which helium would leak from the vessel.

3.7. A 5.66-m^3 vessel containing air at 290 K is to be depressured to the atmosphere in 2 minutes should the pressure ever exceed 10 MPa. Calculate the required diameter for the rupture disk. Isentropic expansion, ideal-gas behavior and a constant heat capacity ratio may be assumed.

3.8. Prepare a plot of the cross-sectional area, velocity, density, mass velocity and the temperature as a function of pressure for 7.6 kg/s of air expanding isentropically from a vessel at 0.7 MPa and 293 K to the atmosphere through a converging-diverging nozzle. At what other pressure could the air discharge isentropically from the same nozzle? How would the mass flow rate change? Prepare the corresponding plot for this new exit condition.

3.9. Plot the mass flow rate and the pressure and velocity at the throat versus upstream pressure for isentropic flow of 288 K air to the atmosphere through a converging nozzle, 100 mm in diameter at the inlet and 50 mm in diameter at the outlet.

3.10. A converging-diverging nozzle is being designed to expand air isentropically from 0.3 MPa to the atmosphere. If the diameter of the throat is to be 50 mm, what should be the diameter at the exit?

3.11. A converging-diverging nozzle is designed properly to expand 13.6 kg/s of air from 0.7 MPa to the atmosphere.
If the upstream pressure is doubled, should the nozzle be extended (or shortened) to increase (or decrease) the exit area? Explain.
If a change is desirable what should be the ratio of the new exit area to the old.

3.12. (a) Calculate the kg/s of air which will flow adiabatically from a vessel at 0.12 MPa and 288 K (ρ = 1.45 kg/m^3) to the atmosphere through a well-designed converging nozzle with an exit diameter of 100 mm.
(b) It is proposed to add a diverging section to the above nozzle. What should be the diameter of the exit?
(c) What flow rate could be achieved if this diverging section were added?
(d) If the diverging section were over-designed, i.e., if the exit diameter were too large, would the flow be greater than, less than or the same as in (c)?
(e) If the diverging section were under-designed, i.e., if the exit diamter were too small would the flow be greater than, less than or the same as in (c)?

3.13. Prepare a plot of the dimensionless flow rates w/w_1 and u_t/u_{t1} versus dimensionless external pressure P_∞/P_0 for a converging-diverging nozzle 50 mm diameter at the throat and 100 mm at the exit, discharging air from a vessel at P_0 = 0.7 MPa. Here the subscript 1 indicates the flow rates which are obtained when exhausting to atmospheric pressure.

3.14. (a) Plot the variation of the mass flow rate with pressure as a vessel containing air at 0.7 MPa and 288 K is depressured to the atmosphere through a converging-diverging nozzle with a throat area of 710 mm^2 and an exit area of 1225 mm^2.
(b) Calculate the pressure in the vessel at the time when the flow through the throat of the nozzle ceases to be sonic.

3.15. Air expands isentropically through a nozzle from a reservoir at 310 K. What is the Mach number at the point in the nozzle where the temperature is 275 K?

3.16. Is it possible to have sonic velocity in the throat of a converging-diverging nozzle through which air is discharged to the atmosphere (0.1 MPa) from a vessel at 288 K and 0.15 MPa? Adiabatic frictionless flow may be assumed. If your answer is "No", what vessel pressure would be required? If your answer is "Yes", what should be the ratio of the diameter of the throat to the diameter of the outlet of the nozzle?

3.17. 0.252 kg/s of hydrogen at 289 K are discharging to the atmosphere from a 52.5-mm vent line through a diverging nozzle. Because of danger of ignition of the hydrogen–air mixture outside the vent by a shock wave, it is important to avoid *over-expansion.* Does *over-expansion* occur under these conditions?

3.18. If the pressure at the exit of a converging-diverging nozzle lies between the two design pressures, the gas must first expand from a sonic condition to a supersonic condition downstream from the throat, and then pass through a shock front inside the nozzle in order to attain the exit pressure. Compare qualitatively the temperature of the gas leaving the nozzle with the inlet temperature and the temperatures which would be attained at the two design pressures. Explain.

3.19. A vessel containing air at 288 and 0.9 MPa is to be vented to the atmosphere through a 50-mm I.D. flange. An engineer proposes to install a diverging nozzle on the flange to reduce the noise and to increase the mass rate of venting, as shown in Fig. 3.5.

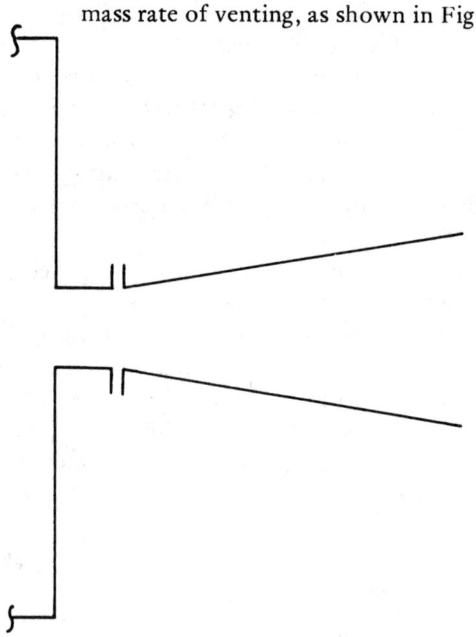

Fig. 3.5. Flanged diverging nozzle

(a) Will the nozzle reduce the noise by eliminating the the possibility of a sudden change in pressure? Explain.

(b) Will the nozzle increase the rate of venting? Explain.

(c) If the answer to either (a) or (b) is "Yes", what should be the diameter of the exit from the nozzle?

3.20. Under what conditions, if any, will the addition of a diverging section to a converging nozzle yield a higher or lower mass flow rate? Explain briefly.

3.21. A vessel is to be depressurized in an emergency through a round hole created by the failure of a rupture disk. Under the influence of this course an engineer proposes to increase the rate of depressurization through the port by installing a converging section inside and a diverging section outside the vessel. Will either or both additions help? Explain briefly.

3.22. Under what conditions, if any, will a converging nozzle yield a higher or lower mass flow rate than a sharp-edged orifice with the same diameter as the throat of the nozzle? Explain briefly.

3.23. A gas is expanded supersonically through a converging-diverging nozzle. How will the mass rate of flow change if

(a) the upstream pressure is doubled?

(b) the upstream temperature is halved?

(c) the downstream pressure is doubled?

(d) the downstream pressure is halved?

3.24. The velocity of sound of a gas is measured to be 207 m/s at 298 K and the molar mass is determined to be 71.23 g/mol. Estimate γ, c_P and c_V.

3.25. Calculate the ratio of critical mass flow through a nozzle for isothermal and isentropic conditions with $\gamma = 1.4$.

3.26. Does friction increase or decrease the mass flow rate in a converging-diverging nozzle relative to the idealized value for isentropic flow? Explain.

3.27. Prepare a quantitative dimensionless plot of mass flow rate and exit temperature versus vessel pressure for air at 294 K venting to the atmosphere through a well-designed converging-diverging nozzle with a throat diameter of 50 mm and an exit diameter of 75 mm.

3.28. Derive expressions for isothermal flow through a converging-diverging nozzle, analogous to equations (3.16) – (3.25) for isentropic flow.

3.29. (a) Show that the maximum possible velocity which can be attained through isentropic expansion of an ideal gas is $a_1\,[2/(\gamma-1)]^{1/2}$ where a_1 is the acoustic velocity at the initial conditions.

(b) Show that the maximum attainable velocity in a converging nozzle is $[(\gamma-1)/(\gamma+1)]^{1/2}$ times the maximum possible velocity.

(c) Show that the maximum attainable velocity in a converging-diverging nozzle is equal to the factor:

$$\left[1 - \left(\frac{P_2}{P_1}\right)^{\frac{\gamma-1}{\gamma}}\right]^{\frac{1}{2}}$$

times the maximum possible velocity for isentropic expansion.

3.30. A stream of air with a velocity of 600 m/s and a temperature of 288 K is to be produced by expansion from a large vessel to the atmosphere through a converging-diverging nozzle.
(a) What temperature is required in the vessel?
(b) What pressure is required in the vessel?
(c) What ratio of throat area to exit area is required?
(d) What will be the pressure at the throat?
(e) What will be the temperature at the throat?
(f) What will be the velocity at the throat?

3.31. Derive Equation (3.5) from the general equations of continuity and motion. Hint: Look up the general equations of motion and reduce for steady, one-dimensional, horizontal, inviscid motion or reduce Equations (1.1) and (1.4). Eliminate du. Reexpress in terms of the frame of reference of a standing wave.

3.32. Bird et al. [10] give the following expression for the time to depressure a tank of volume V through a nozzle. Is this equation correct?

$$t = \frac{V/A_t}{\left[\gamma\left(\frac{P_i}{\rho_i}\right)\left(\frac{2}{\gamma+1}\right)^{\frac{\gamma+1}{\gamma-1}}\right]^{1/2}}\left(\frac{2\gamma}{\gamma-1}\right)\left[1 - \left(\frac{\rho_1}{\rho_i}\right)^{\frac{\gamma-1}{2\gamma}}\right]$$

$$\frac{P_\infty}{P_1} < \left(\frac{2}{\gamma+1}\right)^{\frac{\gamma}{\gamma-1}}$$

Here the subscript i indicates the initial pressure.

3.33. Show that

$$\left(\frac{\partial P}{\partial \rho}\right)_S = \frac{1}{\left(\frac{\partial \rho}{\partial P}\right)_T - \frac{T}{\rho c_P}\left(\frac{\partial \rho}{\partial T}\right)_P^2}$$

3.34. Compare the theoretical value of sound with the following experimental values corrected to 273.15 K:

	m/s		m/s
Air	331.7	Ethylene	314
Acetylene	327.5	Helium	965
Ammonia	415	Hydrogen	1269.5
Argon	319	Methane	432
Carbon dioxide	258	Neon	435
Carbon monoxide	337.1	Nitrogen	334
Chlorine	205.3	Sulfur dioxide	209.2
Ethane	303	Water vapor	401

References

1. R. S. Westfall, "Newton and the Fudge Factor," *Science* **179**, No. 4075, 751-758 (February 23, 1973).
2. G. E. Gorring, "Compact Sonic Velocity Equation for Nonideal Gases," *Ind. Eng. Chem. Fundam.* 7, 660-670 (1968).
3. W. Squire, "Speed of Sound in a Nonideal Gas," *Can. J. Chem. Eng.* **39**, 199-200 (1961).
4. W. B. Whiting, II and R. C. Ackerberg, "Prediction of Acoustic Velocities in Nonideal Gaseous Mixtures," *AIChE Journal* **24**, 341-343 (1978); **25**, 560 (1979).
5. E. Mach, *Sitzber. Wiener Akad. Wiss.* (1887), p. 164; and (1896), p. 605.
6. A. M. Binney and M. W. Woods, "The Pressure Distribution in a Convergent Steam Nozzle," *Proc. Instn. Mech. Eng.* **138**, 259-266 (1938).
7. H. W. Emmons, Ed., *Fundamentals of Gas Dynamics,* Princeton University Press, Princeton, New Jersey (1958), pp. 142-158.
8. H. P. Grace and C. E. Lapple, "Discharge Coefficients of Small-Diameter Orifices and Flow Nozzles," *Trans. ASME* 73, 639-647 (1951).
9. P. W. Godt, "Experimental Correlation of Air, Nitrogen, Helium and Argon Flowrates through Critical Flow Nozzles," *Symposium for Flow–Its Measurement and Control in Science and Industry,* Ed., R. B. Dowdell, Instr. Soc. Amer. (1974), Vol. 3, pp. 263-267.
10. R. B. Bird, W. E. Stewart and E. N. Lightfoot, *Transport Phenomena,* Wiley, New York (1960), pp. 480-482.

Chapter 4

Jet propulsion engines

The design and performance of ramjets and turbojets are complicated by friction, compressors, turbines, combustion, and the details of the internal and external flow patterns. Rocket design and performance are correspondingly complicated. However, the overall performance of these propulsive devices can be interpreted and approximated by direct application of the expressions derived in the previous chapters if the inlet and outlet conditions or their equivalent are specified. Such applications are described in this chapter.

Jet engines

Jet engine operation is illustrated schematically in Fig. 4.1. Air enters, compresses, and oxidizes a liquid fuel. The burned gases exhaust to the atmosphere through a nozzle.

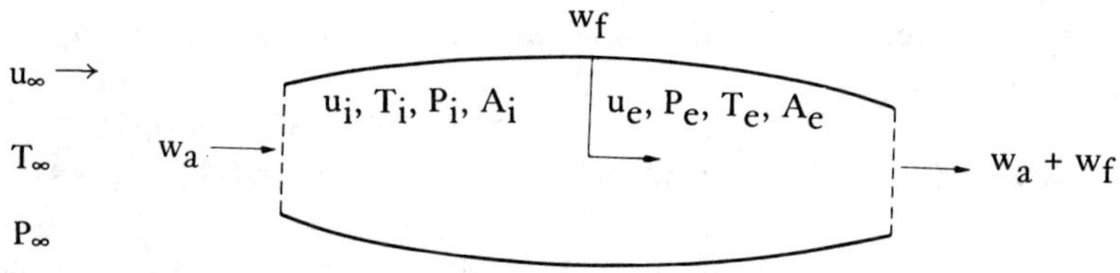

Fig. 4.1. Jet engine operation

With ramjets the air is compressed by decreasing the velocity. In supersonic flight the air is first compressed in passing through a standing shock wave and then in flow through a divergent "subsonic diffusion" chamber as illustrated in Fig. 4.2. In subsonic flight the compression must be accomplished wholly in the diffusion chamber, and the maximum internal pressure is less than the stagnation pressure of the ambient air relative to the moving ramjet. Ramjets are thus efficient only at high velocities.

In turbojets, as indicated in Fig. 4.3, the air is compressed by an axial compressor which is driven by a turbine through which the burned gases pass before entering the exit nozzle. As indicated in Fig. 4.4, some of the fuel may be added in an afterburner following the turbine in order to minimize the temperature of the gases in the turbine.

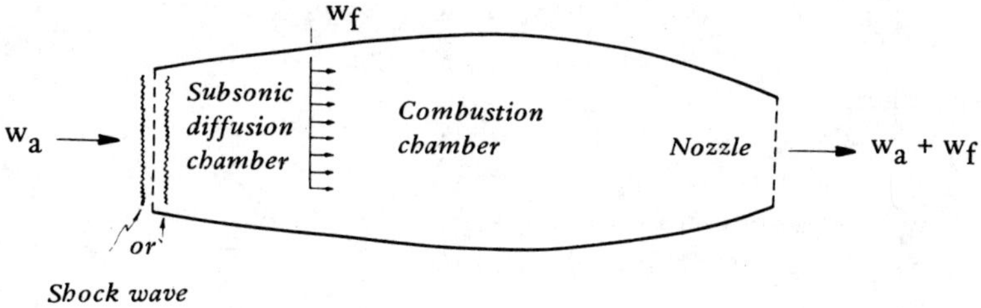

Fig. 4.2. Ramjet engine

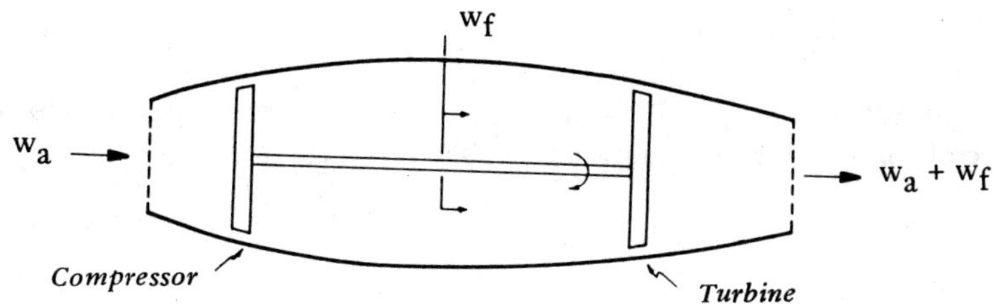

Fig. 4.3. Turbojet engine

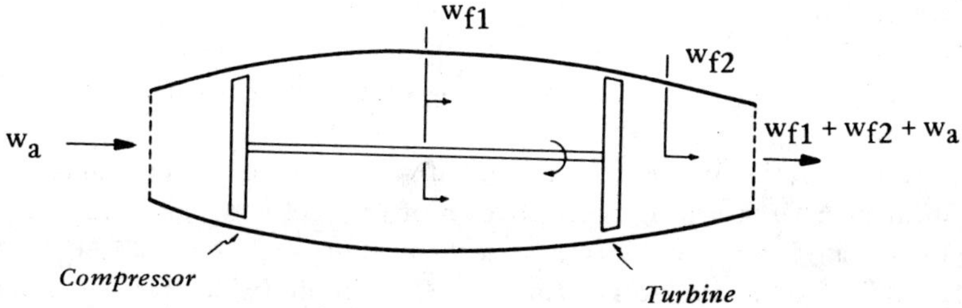

Fig. 4.4. Turbojet engine with afterburner

a) Thrust. The net thrust of jet engines is obtained from an overall force and momentum balance as follows:

$$T = (w_a + w_f)u_e + P_eA_e - w_au_i - P_iA_i + P_\infty(A_i - A_e) \qquad (4.1)$$

where here

T = net thrust of engine, N or m·kg/s^2
w_a = mass rate of flow of air, kg/s
w_f = mass rate of flow of fuel, kg/s
u_e = exhaust velocity, m/s
u_i = inlet velocity, m/s
P_e = pressure on the exhaust area, Pa
P_i = pressure on the inlet area, Pa
P_∞ = ambient pressure, Pa
A_e = exhaust area, m^2
A_i = inlet area, m^2

Ideally $P_e = P_i = P_\infty$ and $u_i = u_\infty$, where here u_∞ is the velocity of the aircraft, in which case Equation (4.1) simplifies to

$$T = (w_a + w_f)u_e - w_au_\infty \qquad (4.2)$$

The velocities in Equation (4.2) can be replaced by mass flow rates and areas through Equation (1.6), and the densities can be eliminated through the ideal gas law with $T_i = T_\infty$, to give

$$T = (w_a + w_f)^2 \frac{R\,T_e}{A_e m_e P_\infty} - \frac{w_a^2 R\,T_\infty}{A_i m_a P_\infty} = \frac{w_a^2 R T_e}{A_e P_\infty m_e}\left((1+\phi)^2 - \frac{A_e T_\infty m_e}{A_i T_e m_a}\right) \qquad (4.3)$$

where here

$$\phi = w_f/w_a = \text{mass ratio of fuel to air}$$

The mass ratio of fuel to air is usually less than 0.06, and Equation (4.3) indicates that the primary contribution of the fuel is to raise the temperature of the exhaust gases by the release of the heat of combustion. Clearly, a low molar mass m_e for the products of combustion is also desirable.

If the conditions ahead of the jet nozzle are known instead of at the exit, the exit conditions can be estimated by assuming isentropic expansion. Jet engines may have converging-diverging nozzles or merely converging nozzles, and the velocity at the throat may be acoustic or less depending on the degree of compression which is attained. Because of the relatively low degree of compression, converging-only nozzles are usually used. If a converging-diverging nozzle is used, drag and other considerations usually dictate an exit area somewhat smaller than required for expansion to P_∞, particularly at high altitudes where $P_\infty \to 0$. In such cases Equation (4.1) reduces to

$$T = w_a u_e \left(1 + \phi - \frac{u_\infty}{u_e}\right) + (P_e - P_\infty)A_e \tag{4.4}$$

and through Equations (1.6) and (1.12) to

$$T = \frac{w_a^2 R T_e}{A_e P_\infty m_e}\left((1 + \phi)^2 - \frac{A_e T_\infty m_e}{A_i T_e m_a}\right) + (P_e - P_\infty)A_e \tag{4.5}$$

rather than to Equations (4.2) and (4.3).

The expressions above are all for flight conditions with $u_i \cong u_\infty$. For stationary tests a converging nozzle is usually attached to the inlet to produce the desired inlet conditions. The inlet conditions then correspond to isentropic expansion from $u_\infty = 0$, T_∞ and P_∞ to u_i, T_i and P_i. P_∞ and T_∞ thus correspond to the stagnation temperature and pressure for T_i and P_i and the latter values can be calculated from Equations (1.45) and (1.38) as follows:

$$T_i = T_\infty - (u_i^2 / 2c_p) \tag{4.6}$$

and

$$P_i = \frac{P_\infty}{1 + \left(\dfrac{u_i^2}{2c_p T_i}\right)^{\frac{\gamma}{\gamma - 1}}} = \frac{P_\infty}{1 - \left(\dfrac{u_i^2}{2c_p T_\infty}\right)^{\frac{\gamma}{\gamma - 1}}} \tag{4.7}$$

Only the component of the exit velocity in the direction of flight contributes to the thrust. Malina [1] reported that Tsien integrated this component across divergent nozzles and arrived at the expression

$$u_e^* = 0.5(1 + \cos\{\theta\})u_e \tag{4.8}$$

where

θ = total angle of a diverging nozzle, rad
u_e^* = effective exit velocity, m/s

b) Efficiencies. A number of different efficiencies are used for comparison of the performance of jet engines. The *propulsion efficiency* of a jet engine may be defined as

$$\eta_J = \frac{T\,u_\infty}{\frac{w_a u_e^2}{2}\left[(1+\phi)^2 - \left(\frac{u_i}{u_e}\right)^2\right]} \tag{4.9}$$

For the idealized conditions of Equation (4.2), Equation (4.9) can be reduced to

$$\eta_J = \frac{2}{1 + (1+\phi)\frac{u_e}{u_\infty}} \tag{4.10}$$

It is apparent from Equation (4.10) that η_J decreases as u_e increases.

The *overall efficiency* of a jet engine may be defined as

$$\eta_0 = \frac{T u_\infty}{Q_l w_f} \tag{4.11}$$

where

Q_l = lower heating value of the fuel, J/kg or m^2/s^2

For the idealized conditions of Equation (4.2)

$$\eta_0 = \frac{u_\infty[(1+\phi)u_e - u_\infty]}{Q_l \phi} \tag{4.12}$$

Rocket engines

A liquid-liquid rocket engine ordinarily operates as indicated in Figure 4.5 with self-contained liquid fuel and liquid oxidant pumped at controlled rates to a reaction chamber operating at elevated pressure. The products of reaction exit through a converging-diverging nozzle. The conditions in the reaction chamber and at the exit are identified by the subscripts 1 and e, respectively.

P_∞ is again the ambient pressure.

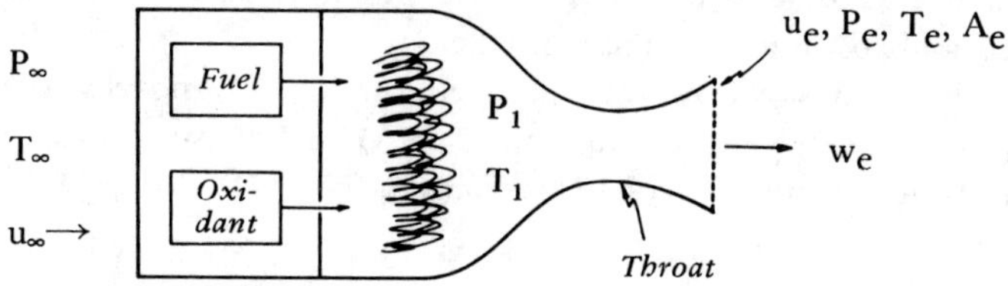

Fig. 4.5. Liquid-liquid-fueled rocket engine

The thrust of such a rocket engine can be expressed as

$$T = w_e u_e + (P_e - P_\infty)A_e \tag{4.13}$$

where here

w_e = mass rate of flow through the nozzle, kg/s

Ordinarily the reaction in a rocket is maintained at a sufficient rate to produce acoustic velocity at the throat of the nozzle and supersonic velocities in the diverging section. For isentropic flow, w_e is given by Equation (3.23) and u_e by Equation (2.14) with P_e substituted for the general P. Thus

$$\frac{T}{A_t P_1} = \gamma \left[\left(\frac{2}{\gamma - 1}\right)\left(\frac{2}{\gamma + 1}\right)^{\frac{\gamma+1}{\gamma-1}} \left(1 - \left(\frac{P_e}{P_1}\right)^{\frac{\gamma-1}{\gamma}}\right)\right]^{1/2} + \left(\frac{P_e}{P_1} - \frac{P_\infty}{P_1}\right)\frac{A_e}{A_t} \tag{4.14}$$

The pressure ratio P_e/P_1 can be related to the area ratio A_e/A_t by substituting w from Equation (3.23) in Equation (3.16) and identifying the general area A and pressure P as A_e and P_e. The result is

$$\left(\frac{A_t}{A_e}\right)^2 = 2\left(\frac{\gamma+1}{2}\right)^{\frac{\gamma+1}{\gamma-1}} \left[\left(\frac{P_e}{P_1}\right)^{\frac{2}{\gamma}} - \left(\frac{P_e}{P_1}\right)^{\frac{\gamma+1}{\gamma}}\right] \tag{4.15}$$

The relationship between thrust and A_e (or P_e), as provided by Equations (4.14) and (4.15), is quite complex and is explored numerically in Problems 4.3 and 4.4. In any event the optimal exit area for maximal thrust would generally require a nozzle of excessive length. The optimal exit area taking into account drag and cost is usually much smaller.

For acoustic flow through a converging-only nozzle $A_e = A_t$ and $P_e = P_t$. P_1/P_e is then given by Equation (3.18A) with P_e substituted for P_c. With these substitutions Equation (4.13) reduces to

$$\frac{T}{A_t P_1} = 2\left(\frac{2}{\gamma + 1}\right)^{\frac{1}{\gamma - 1}} - \frac{P_\infty}{P_1} \tag{4.16}$$

Equation (4.16) can also be derived directly from Equation (4.13) with w_c, u_c, and P_c substituted from Equations (3.23), (3.21), and (3.18A) for w_e, u_e, and P_e.

Rocket engines and propellants are often compared in terms of their *specific impulse* defined as the thrust divided by the mass flow rate:

$$I \equiv T/w_e \tag{4.17}$$

For propellants this comparison is usually made for $P_e = P_\infty$ and for $P_\infty = 1$ atm or zero. Equations (4.14) and (3.23) give the following expression for the maximum specific impulse, assuming isentropic expansion of an ideal gas with constant heat capacity to $P_e = 0$:

$$I_{max} = \left[\left(\frac{2\gamma}{\gamma - 1}\right)\frac{P_1}{\rho_1}\right]^{1/2} = \left[\left(\frac{2\gamma}{\gamma - 1}\right)\frac{R\,T_1}{m_1}\right]^{1/2} \tag{4.18}$$

I has the dimensions of m/s or N·s/kg. It is apparent that the specific impulse for the conditions of Equation (4.18) is identically the maximum limiting velocity obtained by isentropic expansion to zero pressure. The American rocket industry usually expresses specific impulse in lb_f·s/lb and then often erroneously abbreviates this as *seconds.*

Equation (4.18) indicates that in order to have a high specific impulse the reactants should produce gaseous products with a low molecular weight and a low γ at very high temperature. Glassman [2] discusses the maximum possible specific impulse which can be obtained with various chemical combinations.

Rocket chamber pressure

The mass rate of generation of gaseous products of reaction must equal the mass rate of flow out the rocket nozzle during stable operation. Liquid fuel and liquid oxidant are generally fed at controlled rates. Equating the combined rate of feed to the rate of flow as given by Equation (3.23) with the density replaced through the ideal gas law:

$$w_e = A_t P_1 \left[\frac{\gamma m_1}{R T_1} \left(\frac{2}{\gamma + 1} \right)^{\frac{\gamma + 1}{\gamma - 1}} \right]^{1/2} \qquad (4.19)$$

Hence, the reaction chamber pressure is

$$P_1 = \frac{w_e}{A_t} \left[\frac{R T_1}{\gamma m_1} \left(\frac{\gamma + 1}{2} \right)^{\frac{\gamma + 1}{\gamma - 1}} \right]^{1/2} \qquad (4.19A)$$

The situation is somewhat more complicated with a rocket driven by a solid propellant as illustrated in Figure 4.6. The reaction will be presumed to convert

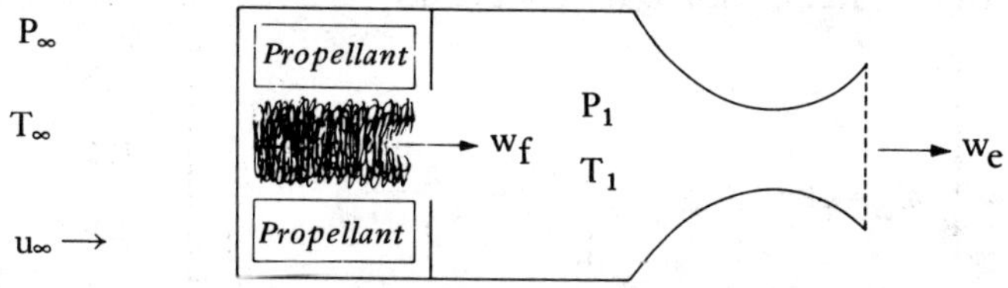

Fig. 4.6. Solid-fueled rocket engine

the solid completely to gas. The mass rate of production of gas by reaction can be expressed as

$$w_f = \rho_p A_f \bar{r} \qquad (4.20)$$

where

ρ_p = density of solid propellant, kg/m^3
A_f = surface area for reaction, m^2
$\bar{r}$ = linear rate of gasification of solid propellant, m/s

In general $\bar{r}$ will be a function of both temperature and pressure. The dependence on temperature will be neglected herein and the dependence on pressure postulated to be a power function such that

$$\tilde{r} = bP_1^n \tag{4.21}$$

where b and n are constants.

The rate of change of pressure in the reaction chamber can be expressed in terms of the rates of reaction and venting as follows:

$$\frac{dP_1}{dt} = \frac{(w_f - w_e)RT_1}{V_1 m_1} \tag{4.22}$$

where

V_1 = gas space in reaction chamber, m^3
t = time, s

Substituting for w_f from Equation (4.20), for $\tilde{r}$ from Equation (4.21) and for w_e from Equation (3.23), with again the density replaced through the ideal gas law, gives

$$\frac{dP_1}{dt} = \left[b\rho_p A_f P_1^n - A_t P_1 \left[\frac{\gamma m_1}{RT_1} \left(\frac{2}{\gamma+1} \right)^{\frac{\gamma+1}{\gamma-1}} \right] \right] \frac{RT_1}{V_1 m_1} \tag{4.23}$$

Equation (4.23) indicates that if n = 1, P_1 will increase exponentially and indefinitely with time. Hence the propellant will explode and the rocket chamber will blow up. If $n > 1$ the pressure will increase even more rapidly and disasterously. If $n \ll 1$ the pressure will rise to

$$P_1 = \left[\frac{b\rho_p A_f}{A_t} \left(\frac{RT_1}{\gamma m_1} \left(\frac{\gamma+1}{2} \right)^{\frac{\gamma+1}{\gamma-1}} \right)^{1/2} \right]^{\frac{1}{1-n}} \tag{4.24}$$

and remain essentially constant until the propellant is completely reacted. Of course, some variation in pressure can be expected due to secondary effects, such as the changing surface area for reaction and the slow increase in volume of the reaction chamber available to the gas.

Solid propellants have advantages in terms of storage but are clearly more difficult to control than liquid propellants. Hybrid propellants, in which a liquid or gaseous oxidant (or fuel) is fed at a controlled rate to the surface of a solid fuel (or oxidant), as illustrated in Fig. 4.7, combine some of the advantages of completely solid and completely liquid propellants.

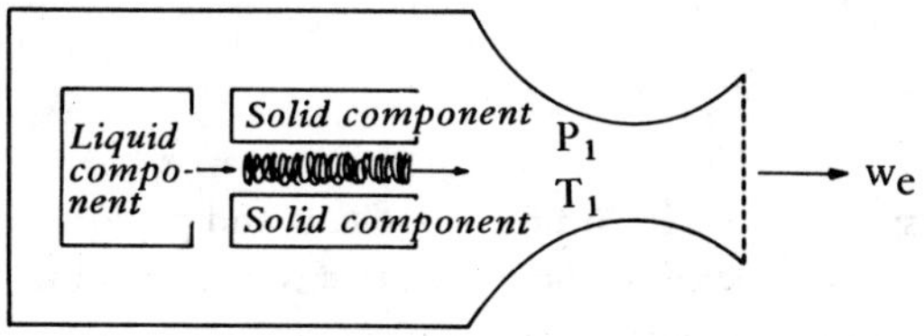

Fig. 4.7. Hybrid-propellant rocket engine

Rocket dynamics

The previous expressions for rocket motors were derived implicitly with the rocket as the frame of reference for the velocities, i.e., for air moving past a stationary rocket. The stagnant air will now be considered as a frame of reference. These two frames of reference are compared in Figs. 4.8 and 4.9.

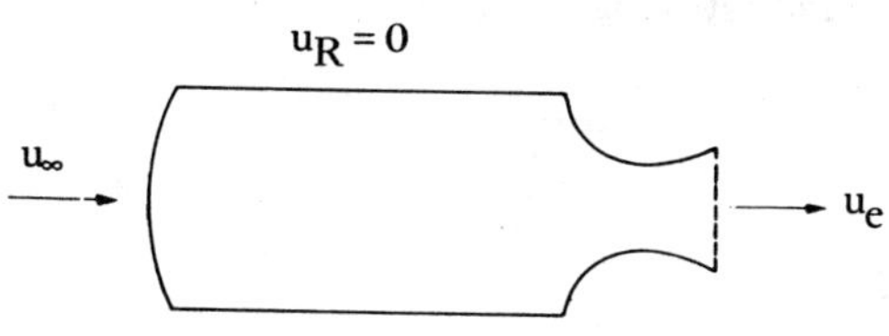

Fig. 4.8. Frame of reference for a stationary rocket

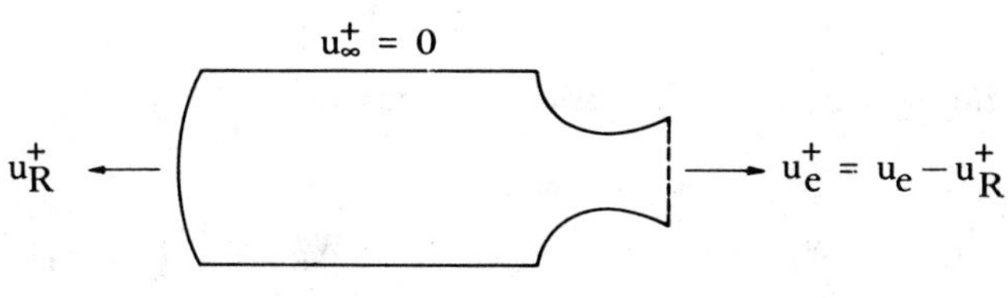

Fig. 4.9. Frame of reference for a moving rocket

The time rate of increase of momentum of the rocket is equal to the rate of momentum transfer to the atmosphere by venting (in the opposite direction), less the gravitational force, less the drag force, or

$$\frac{d(W_R u_R^+)}{dt} = w_e u_e^+ - gW_R \cos\{\psi\} - F_D \tag{4.25}$$

where

W_R = mass of rocket, including propellant and payload, kg
u_R^+ = forward velocity of rocket relative to the surroundings, m/s
u_e^+ = backward velocity of vented gas relative to the surroundings, m/s
F_D = drag force of atmosphere, N or m·kg/s²
ψ = angle of rocket motion relative to the vertical, rad

Equation (4.25) is actually for the component of velocity opposite to the thrust, and implies that the component of gravity in the sidewise direction is negligible.

Expanding the derivative:

$$\frac{d(W_R u_R^+)}{dt} = \frac{W_R \, du_R^+}{dt} + u_R^+ \frac{dW_R}{dt} \tag{4.26}$$

A material balance indicates that

$$-\frac{dW_R}{dt} = w_e \tag{4.27}$$

Also as indicated in Fig. 4.9

$$u_e^+ = u_e - u_R^+ \tag{4.28}$$

Substituting these three expressions in Equation (4.25) gives

$$W_R \frac{du_R^+}{dt} = w_e u_e - gW_R \cos\{\psi\} - F_D \tag{4.29}$$

If w_e is constant, Equation (4.27) can be integrated to give

$$W_R = W_{R0} - w_e t \tag{4.30}$$

where

W_{RO} = initial mass of rocket, including propellant and payload, kg

Neglecting F_D as a first approximation (which is valid outside the atmosphere), substituting for W_R in Equation (4.29) from (4.30), and rearranging:

$$du_R^+ = \left(\frac{w_e u_e}{W_{RO} - w_e t} - g\ cos\{\psi\} \right) dt \tag{4.31}$$

Integrating from $u_R^+ = 0$ at $t = 0$, with u_e and $g\cos\{\psi\}$ assumed constant gives

$$u_R^+ = u_e \ln \left\{ \frac{1}{1 - \frac{w_e t}{W_{RO}}} \right\} - gt\ cos\{\psi\} \tag{4.32}$$

The final or burnout velocity is then

$$(u_R^+)_F = u_e \ln \left\{ \frac{1}{1 - \frac{W_{f0}}{W_{RO}}} \right\} - gt_F\ cos\{\psi\} \tag{4.33}$$

$$= u_e \ln \left\{ \frac{W_{RO}}{W_{RO} - W_{f0}} \right\} - g\ \frac{W_{f0}}{w_e}\ cos\{\psi\} \tag{4.33a}$$

where

W_{f0} = initial mass of propellant, kg

If the atmospheric pressure is negligibly small and if the exit area is very large, u_e^+ can be approximated by I_{max}. If the time of flight is very short or the flight is horizontal, gW_{f0}/w_e may be negligible. Then

$$(u_R^+)_F = I_{max} \ln \left\{ \frac{1}{1 - \frac{W_{f0}}{W_{RO}}} \right\} \tag{4.34}$$

Equation (4.34) provides an upper limit on the velocity which can be obtained from a single-stage, chemical rocket.

Supplementary reading

A very elementary and general treatment of rocketry is provided by Ley [3]. More detailed and technical treatments are given by Lewis [4] and Lancaster [5]. An excellent history of the development of jet propulsion devices beginning with the Greeks in the fourth century B.C. is included in the latter volume.

Problems

4.1. A jet engine produces a thrust of 53 kN when operated on a test stand with ambient air entering at 243 m/s and exhaust gases leaving at 488 m/s. The fuel/air ratio is 0.05. Calculate the fuel consumption rate, the propulsion efficiency, and the overall efficiency assuming Q_l = 44 kJ/kg.

4.2. Calculate the thrust and efficiencies which will be achieved at a velocity of 240 m/s and an altitude of 11.5 km with the same fuel consumption and fuel-to-air ratio. The ambient temperature and pressure may be assumed to be 218 K and 0.16 MPa.

4.3. Prepare a plot of T/A_tP_1 and P_e/P_1 versus A_e/A_t for $P_e = P_\infty$.

4.4. Prepare a plot of T/A_tP_1 versus A_e/A_t for P_∞/P_t = 0, 0.0005, 0.01, and 0.05. Draw and interpret curves through the maxima.

4.5. Calculate the maximum specific impulse for
   a) a propellant producing gas at 2480 K with m_1 = 0.012 and γ = 1.2.
   b) a solid propellant producing a gas at 2800 K containing equal amounts of H_2O, CO_2, CO and N_2.
   c) a stoichiometric mixture of liquid H_2 and liquid O_2 assuming complete reaction.
   d) a stoichiometric mixture of liquid H_2 and F_2 assuming complete reaction.

4.6. A mixture of 3.2 kg O_2/kg H_2 is reported to produce a temperature of 2755 K, an average molar mass of 0.0086 and a specific impulse at sea level of 3376 N/kg·s when burned at 2 MPa. Estimate the effective value of γ.

4.7. Calculate the thrust per unit throat area and the propulsion efficiency for operation at sea level for a well-designed nozzle with $P_e = P_\infty$ and the propellant in
   a) Problem 4.5 (a) with P_1 = 3 MPa.
   b) Problem 4.5 (b) with P_1 = 1.5 MPa.
   c) Problem 4.5 (c) with P_1 = 2 MPa.
   d) Problem 4.5 (d) with P_1 = 2 MPa.
   e) Problem 4.6.

4.8. A rocket engine is to develop a thrust of 6.67 MN. Calculate the required throat and exit areas and the fuel consumption rate for the conditions of

a) Problem 4.7 (a).
b) Problem 4.7 (b).
c) Problem 4.7 (c).
d) Problem 4.7 (d).
e) Problem 4.7 (e).

4.9. A solid propellant has the properties below. Calculate the steady state pressure attained on burning in a rocket engine with the indicated dimensions.

$b = 0.004422\ ft^2/s \cdot lb_f^{1/2}$
$n = 0.5$
$A_f = 6.0\ ft^2$
$\rho_p = 40\ lb/ft^3$
$T_1 = 4000\ °R$
$\gamma = 1.15$
$m_1 = 0.015$
$A_t = 0.75\ ft^2$

4.10. Calculate the specific impulse and maximum possible thrust at sea level of the rocket engine in Problem 4.9. What exit area would be required?

4.11. Equation (4.23) can be arranged in the form

$$\frac{1}{P_1}\ \frac{dP_1}{dt^*} = b^*\, P_1^{n-1} - 1$$

where t* and b* are groups of variables. Prepare a plot of P_1 versus t* for b* = 1.0 and n = 0, 0.5, 1 and 2.

4.12. Calculate the maximum possible burnout velocity for the propellant in Problem 4.6 assuming 90% of the initial rocket weight is fuel.

4.13. Calculate the maximum possible burnout velocities, assuming $W_{f0}/W_{R0} = 0.9$, for the propellants in

a) Problem 4.5a.
b) Problem 4.5b.
c) Problem 4.5c.
d) Problem 4.5d.
e) Problem 4.5e.

4.14. The top of a cylinder containing H_2 at 500 psia is knocked off. The resulting opening is 3 in. in diameter. The mass of the empty tank is 75 lbs and the gas volume is 2 ft^3. Calculate the force and terminal velocity imparted to the tank. Neglect drag and gravitational forces.

4.15. The length of the diverging nozzle in Problem 4.8b is limited to 8 times the throat diameter and a maximum angle of 7 degrees is allowed. Calculate the resulting percentage decrease in thrust.

4.16. The burned gas in a jet engine is to be expanded from 1920 K and 140 kPa to 35 kPa in a converging nozzle. How much gain in thrust might be expected from the use of a converging-diverging nozzle? The flight speed is 180 m/s.

4.17. Does the extraction of energy from the burned gases to compress the air reduce the thrust? Does the frictional loss in the turbines reduce the thrust? Does heat leakage through the engine wall reduce the thrust? Explain.

4.18. What is the effect of altitude on the thrust of a jet engine and a rocket?

4.19. Assuming isentropic expansion is maintained, calculate the lost thrust due to divergence if a nozzle with a cone angle of 12 degrees is used for the engine in Problem 4.8a when exhausting at sea level. How long would the diverging section be?

4.20. Find the speed of rotation of the lawn sprinkler shown below for a discharge rate of 10 gal/min assuming no frictional resistance to rotation. $D_e = 0.3$ in.

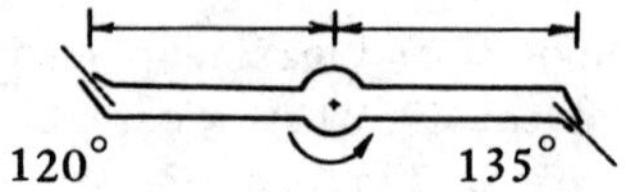

Fig. 4.10. Jet-propelled rotating lawn sprinkler

4.21. Calculate the specific impulse of saturated steam at 7 MPa expanding to the atmosphere through a converging nozzle. Assume condensation does not occur.

4.22. Equation (4.18) indicates that the specific impulse of a propellant is a function of temperature only. Doesn't the chamber pressure affect rocket performance? Explain.

References

1. F. J. Malina, "Characteristics of the Rocket Motor Unit Based on the Theory of Perfect Gases," *J. Franklin Institute* **230**, 433–454 (1040).
2. I. Glassman, "The Chemistry of Propellants," *American Scientist* **53**, 508–524 (1965).
3. W. Ley, *Rockets, Missiles and Space Travel,* Viking, New York (1957).
4. A. D. Lewis, *Gas Power Dynamics,* Van Nostrand, Princeton, New Jersey (1962).
5. O. E. Lancaster, Ed., *Jet Propulsion Engines,* Princeton University Press, Princeton, New Jersey (1959).

Chapter 5

Maximum rate of flow of gas through a pipe

Viscous effects cannot be neglected in flow through a pipe. Nevertheless the rate of flow of a gas is found to be limited by the attainment of acoustic velocity just as in nozzles. This limitation is insiduous and may lead to the gross under-design of piping systems.

Isothermal flow of an ideal gas

The derivation for isothermal flow is somewhat simpler than for adiabatic flow and will therefore be used as an introductory step. The isothermal results also have practical value as a bounding case. Due to significant heat transfer from the surroundings actual behavior may lie between the predictions for isothermal and adiabatic flow. Fortunately, these two solutions do not differ greatly even in a quantitative sense.

For simplicity, horizontal flow will be assumed. The total momentum will be approximated by the mass rate of flow times the mixed-mean velocity (see Problem 5.16). A force and momentum balance can then be written over the element of fluid shown in Figure 5.1 as follows:

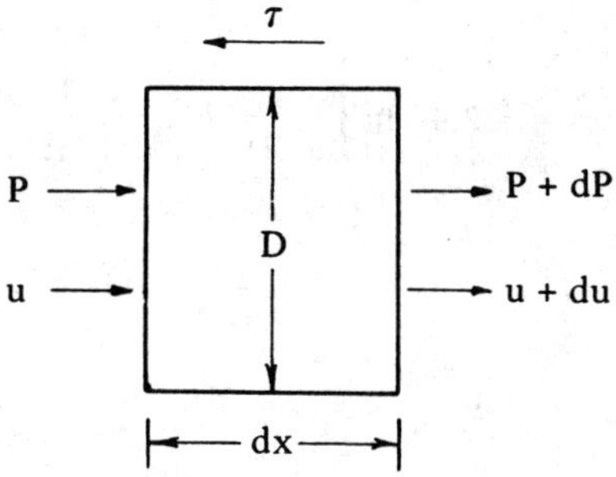

Fig. 5.1. Force balance over a differential disk of fluid in a pipe.

$$\tau\pi D dx + \frac{\pi D^2 dP}{4} + u_b\rho \frac{\pi D^2}{4} du_b = 0 \qquad (5.1)$$

where here

τ = shear stress imposed on fluid by wall or vice versa, Pa or kg/m·s^2 or N/m^2

u_b = mixed-mean or bulk velocity in pipe, $4w/\rho\pi D^2$, m/s

D = pipe diameter, m

The shear stress due to the wall can be expressed in terms of a dimensionless friction factor defined variously as

$$f = \frac{f_{Fanning}}{2} = \frac{f_{Darcy}}{8} = \frac{\tau}{\rho u_b^2} \tag{5.2}$$

All three of these friction factors are widely used. The choice is merely a matter of personal preference. Any one may be substituted for the other via Equation (5.2). Herein the unsubscripted one will be used. In fully developed flow f is a function of Re and e/D

where here

Re = Reynolds number in pipe = $Du_b\rho/\mu = 4w/\pi D\mu = DG/\mu$
e = roughness of pipe, m
G = $u_b\rho = 4w/\pi D^2$ = mass velocity, kg/m^2·s

Turbulent motion can be expected in the limiting case of maximum flow. The following empirical equations of Colebrook [1] for f in the turbulent regime were recently rederived and tested by Churchill [2]:

$$\frac{1}{f^{1/2}} = -2.46 \ln\left\{\frac{e}{3.7D} + \frac{1}{1.13\,Re\,f^{1/2}}\right\} \tag{5.3}$$

and

$$\frac{1}{f^{1/2}} = -2.46 \ln\left\{\frac{e}{3.7D} + \left(\frac{7}{Re}\right)^{0.9}\right\} \tag{5.4}$$

The former is more convenient if the flow rate is to be determined for a specified shear stress and the latter if the shear stress is to be determined for a specific flow rate. The dependence on viscosity decreases as the roughness and the rate of flow increase.

w and G are constant down the pipe. The variation of the viscosity with pressure is ordinarily negligible. Hence $Re = DG/\mu$ is nearly constant. Since e/D is constant, f is also nearly constant down the pipe.

Substituting G/ρ for u_b, and $f = \tau/\rho u_b^2$ in Equation (5.1) and rearranging:

$$f\,dx = -\frac{D\rho}{4G^2}\,dP + \frac{D}{4}\frac{d\rho}{\rho} \qquad (5.5)$$

Substituting for ρ from the ideal gas law:

$$f\,dx = -\frac{DmP}{4G^2RT}\,dP + \frac{D}{4}\frac{dP}{P} \qquad (5.6)$$

Integrating from $x = 0$ and $P = P_1$ to $x = L$ and $P = P_2$ for constant f and rearranging:

$$\frac{G^2}{P_1\rho_1} = \frac{1 - \left(\frac{P_2}{P_1}\right)^2}{\frac{8fL}{D} + 2\ln\left\{\left(\frac{P_1}{P_2}\right)\right\}} \qquad (5.7)$$

According to Equation (5.7), G approaches zero as $P_2 \rightarrow P_1$ and as $P_2 \rightarrow 0$, and is positive for $0 < P_2 < P_1$. Hence a maximum value of G exists for some P_2. This value can be found by equating the derivative of $G^2/P_1\rho_1$ with P_2/P_1 to zero as below. (f is assumed to be constant in this differentiation since Equations (5.3) and (5.4) indicate that the variation of f with Re is slight.)

$$\frac{d(G^2/P_1\rho_1)}{d(P_2/P_1)} = 0 = \frac{-2\left(\frac{P_2}{P_1}\right)\left[\frac{8fL}{D} + \ln\left\{\left(\frac{P_1}{P_2}\right)^2\right\}\right] + 2\left(\frac{P_1}{P_2}\right)\left[1 - \left(\frac{P_2}{P_1}\right)^2\right]}{\left[\frac{8fL}{D} + \ln\left\{\left(\frac{P_1}{P_2}\right)^2\right\}\right]^2} \qquad (5.8)$$

Rearranging:

$$\left(\frac{P_1}{P_{2c}}\right)^2 - \ln\left\{\left(\frac{P_1}{P_{2c}}\right)^2\right\} = \frac{8fL}{D} + 1 \qquad (5.9)$$

Here the subscript c designates the critical value of the variables for which G is a maximum. Eliminating 8fL/D between Equations (5.7) and (5.9) gives

$$\frac{G_c^2}{P_1\rho_1} = \frac{1-\left(\frac{P_{2c}}{P_1}\right)^2}{\left(\frac{P_1}{P_{2c}}\right)^2 - 1} = \left(\frac{P_{2c}}{P_1}\right)^2 \quad (5.10)$$

Rewriting in terms of u_{b2c} and eliminating P_1 and ρ_1:

$$u_{b2c}^2 = \left(\frac{G_c}{\rho_{2c}}\right)^2 = \left(\frac{P_{2c}}{\rho_{2c}}\right)^2 \frac{\rho_1}{P_1} = \frac{P_{2c}}{\rho_{2c}} \quad (5.11)$$

Thus the maximum flow rate corresponds to isothermal acoustic velocity at the exit of the pipe!

Substituting for P_{2c}/P_1 in Equation (5.9) from (5.10) gives

$$\frac{P_1\rho_1}{G_c^2} = 1 + \frac{8fL}{D} + \ln\left\{\frac{P_1\rho_1}{G_c^2}\right\} \quad (5.12)$$

which can be used for design calculations as follows: postulate a value of G, calculate $Re = DG/\mu$, calculate f from Equation (5.4), then calculate G from Equation (5.12) and repeat the entire process as required.

Equations (5.9) – (5.12) are valid only if the pressure outside the exit of the pipe, P_∞, is less than P_{2c}. If $P_{2c} < P_\infty$, the flow rate will be less than G_c and can be calculated from Equation (5.7) with P_∞ substituted for P_2. A test calculation of P_{2c} is necessary to determine which value of G and P_2 will occur.

If acoustic velocity occurs in a pipe it necessarily occurs at the outlet. Otherwise the velocity would be greater than acoustic at the exit.

Adiabatic horizontal flow of an ideal gas

The momentum will again be approximated by wu. Hence Equation (5.5) is applicable.

For adiabatic, free flow of an ideal gas with no heat capacity, equation (1.43) follows from (1.16), (1.17), (1.23) and (1.30) and hence holds for irreversible as well as reversible flow. (See Problem 1.1). Replacing u by G/ρ and noting that

$$c_p[T - T_1] = \frac{c_p m}{R}\left(\frac{P}{\rho} - \frac{P_1}{\rho_1}\right) = \frac{\gamma}{\gamma - 1}\left(\frac{P}{\rho} - \frac{P_1}{\rho_1}\right) \tag{5.13}$$

gives

$$\frac{G^2}{2}\left(\frac{1}{\rho^2} - \frac{1}{\rho_1^2}\right) = \frac{\gamma}{\gamma - 1}\left(\frac{P_1}{\rho_1} - \frac{P}{\rho}\right) \tag{5.14}$$

Solving for P and differentiating with respect to ρ gives

$$dP = \left[\frac{G^2}{2}\left(\frac{\gamma - 1}{\gamma}\right)\left(\frac{1}{\rho_1^2} + \frac{1}{\rho^2}\right) + \frac{P_1}{\rho_1}\right] d\rho \tag{5.15}$$

Substituting this expression for dP in Equation (5.5), and then simplifying and rearranging gives

$$\frac{8f\,dx}{D} = \left(\frac{\gamma + 1}{\gamma}\right)\frac{d\rho}{\rho} - \frac{2}{\rho_1^2}\left(\frac{\gamma - 1}{2\gamma} + \frac{P_1\rho_1}{G^2}\right)\rho d\rho \tag{5.16}$$

Assuming constant f and integrating from ρ_1 at x = 0 to ρ_2 at x = L gives

$$\frac{8fL}{D} = \left(\frac{\gamma + 1}{\gamma}\right)\ln\left\{\frac{\rho_2}{\rho_1}\right\} + \left(1 - \left(\frac{\rho_2}{\rho_1}\right)^2\right)\left(\frac{\gamma - 1}{2\gamma} + \frac{P_1\rho_1}{G^2}\right) \tag{5.17}$$

The temperature will of course vary with x for adiabatic, irreversible flow. Hence μ and DG/μ will vary. However, the resulting variation of f will be negligible under most circumstances.

From Equation (5.17) it can be deduced that $G \to 0$ when $\rho_2 \to \rho_1$ and when $\rho_2 \to 0$, and is positive when $0 < \rho_2 < \rho_1$. Hence a maximum value exists for some ρ_2 and can be found by equating the derivative of $G^2/P_1\rho_1$ with respect to $(\rho_2/\rho_1)^2$ to zero. f is again assumed to change negligibly with G. This procedure gives

$$\frac{P_1\rho_1}{G_c^2} = \frac{\gamma + 1}{2\gamma}\left(\frac{\rho_1}{\rho_2}\right)^2 - \frac{\gamma - 1}{2\gamma} \tag{5.18}$$

Substituting for P_1 from Equation (5.14) with P and ρ identified as ρ_{2c} and P_{2c} gives

$$G_c^2 = \gamma P_{2c} \rho_{2c} \tag{5.19}$$

or

$$u_{b2c}^2 = \gamma P_{2c}/\rho_{2c} \tag{5.20}$$

which is the isentropic acoustic velocity. Thus the criterion for maximum flow in a pipe under either adiabatic or isothermal conditions is the attainment of acoustic velocity at the exit. The irreversibilities simply increase the required pressure ratio.

Substituting for ρ_1/ρ_2 in Equation (5.17) from (5.18) and identifying G as G_c gives on rearrangement

$$\frac{P_1\rho_1}{G_c^2} = \frac{8fL}{D} + \frac{1}{\gamma} + \frac{\gamma+1}{2\gamma} \ln\left\{\frac{\gamma-1}{\gamma+1} + \frac{2\gamma}{\gamma+1}\,\frac{P_1\rho_1}{G_c^2}\right\} \tag{5.21}$$

which is the design equation. Again a value of G_c can be postulated, Re calculated, f determined from Equation (5.4), and G_c calculated from Equation (5.21). This process is repeated as necessary. ρ_2/ρ_1 is then calculated from Equation (5.18) and P_{2c} from Equation (5.19). If $P_{2c} < P_\infty$ sonic flow will not be attained, $P_2 = P_\infty$ and G is obtained by the trial and error solution of Equations (5.3), (5.14) and (5.17). (See Problem 5.14).

Inspection of Equations (5.12) and (5.21) indicates that

$$\frac{G_c^2}{P_1\rho_1} \cong \frac{D}{8fL} \tag{5.22}$$

for both isothermal and adiabatic flow. Therefore these two cases do not differ significantly. Equation (5.22) is a good first approximation for the trial and error solution of Equations (5.12) and (5.21).

Problems

5.1. Calculate the rate of flow of air from a vessel at 0.7 MPa and 193 K to the atmosphere through 30 m of 50 mm pipe with a roughness of 46 μm,
a) assuming adiabatic flow.
b) assuming isothermal flow.
c) using Equation (5.22).

5.2. Ethylene is pumped adiabatically through a horizontal pipe. The inlet conditions are $T_1 = 311$ K and $P_1 = 0.7$ MPa. The exit pressure is such that the maximum possible mass velocity of 245 kg/s·m^2 is just achieved. Estimate the temperature of the gas at the pipe exit.

5.3. Methane at 1.4 MPa enters a 0.3 m I.D. pipeline which is 16 km long. The exit pressure is adjusted to that value which produces the maximum possible mass flow rate. If the flow is essentially isothermal at 15 °C, estimate within 5% the total rate of heat transfer from the atmosphere to the pipe.

5.4. The capacity of a process unit is limited by the rate of flow of a gas stream from one vessel to another through a section of straight pipe. An engineer concludes that critical flow is occurring in the pipe and proposes to increase the flow rate by heating the pipe. Is his recommendation sound? Justify your answer.

5.5. Flow of gas through a long pipeline is found to be limited by sonic velocity at the exit. An engineer suggests that increasing the diameter of the last 20% of the pipeline will remove this restriction. Is his proposal sound? Explain briefly.

5.6. The flow of gas from a vessel at high pressure to one at lower pressure is presumed to be limited by the attainment of sonic conditions in the interconnecting pipe. How much should the diameter of the pipe be increased to double the mass rate of flow?

5.7. Is the flow of water in a pipe ever limited by sonic velocity? Explain briefly.

5.8. If the length of pipe connecting two vessels at fixed pressures is doubled what will happen to the mass flow rate if the flow was initially

a) laminar?

b) turbulent (fully developed)?

c) sonic?

Answer as quantitatively as possible, noting any alternative possibilities.

5.9. The flow of air through a pipe from a vessel at 0.7 MPa to the atmosphere is limited by the attainment of sonic velocity at the exit.

a) Could the mass rate of flow be increased by adding a diverging nozzle on the end of the pipe? Explain.

b) Could the mass rate of flow be decreased by adding a diverging nozzle on the end of the pipe? Explain.

c) Would the velocity at the exit of the nozzles in (a) and (b) be subsonic, sonic or supersonic? Explain.

5.10. Does friction accelerate or decelerate adiabatic flow through a pipe?

5.11. Do the limiting rates of flow for isothermal and adiabatic expansions differ significantly from one another

a) in converging nozzles?

b) in pipes?

Explain briefly.

5.12. Air is flowing from a vessel at 0.3 MPa to one at 0.1 MPa through a section of pipe. Will the flow be limited by the attainment of sonic velocity? Explain.

5.13. The limiting mass rates of flow through a nozzle differ significantly for adiabatic and isothermal conditions but those for a pipe do not. Explain briefly.

5.14. Eliminate (ρ_2/ρ_1) from Equation (5.17) using Equation (5.14), giving an expression which is applicable for subsonic, adiabatic flow.

5.15. Calculate the pressure ratio which will just produce critical adiabatic flow of air through a smooth 52.5 mm pipe, 30 m long.

5.16. Calculate the ratio of the momentum-mean to the mixed-mean velocity in a pipe using the following expressions for the local velocity

a) $$\frac{u}{u_c} = \left(\frac{2y}{D}\right)^{1/7} \tag{5.23}$$

b) $$u\left(\frac{\rho}{\tau}\right)^{1/2} = 5.5 + 2.5 \ln\left\{\frac{y(\tau\rho)^{1/2}}{\mu}\right\} \tag{5.24}$$

where here

u_c = velocity at centerline, m/s

y = distance from tube wall

5.17. Derive the analog of the following expressions for flow between parallel plates separated by distance d:

a) Equations (5.5) – (5.12).

b) Equations (5.16) – (5.21).

What is the definition of f for these expressions?

5.18. Repeat Problem 5.16 for flow in the annulus between concentric pipes of outer diameter D_2 and inner diameter D_1.

References

1. C. F. Colebrook, "Turbulent Flow in Pipes with Particular Reference to the Transition between the Smooth and Rough Pipe Laws," *J. Instn. Civil Eng.* **11**, 133-156 (1938/1939).

2. S. W. Churchill, "Empirical Expressions for the Shear Stress in Turbulent Flow in Commercial Pipes," *AIChE Journal* **19**, 375-376 (1973).

Chapter 6

Sudden expansions and contractions

The contraction of fluid through a sharp-edged orifice or weir was shown in Chapter 2 to be quite reversible. The sudden contraction of a fluid in flowing from a vessel to a pipe, or from a large pipe to a smaller one, initially follows a very similar pattern. On the other hand, the eventual expansion of the fluid from the *vena contracta* to the perimeter of the channel is highly irreversible in all of these cases. The sudden expansion on flow from a pipe to a vessel or larger pipe is also irreversible. Surprisingly, the pressure changes can be estimated from one-dimensional mass and momentum balances. The corresponding irreversibilities can, in turn, be determined from a mechanical energy balance. Derivations of these quantities are presented in the following sections for several simple but important applications.

Mechanical energy balance

For free, horizontal flow (no work) Equation (1.16) reduces to

$$q = \Delta H + \Delta\left(\frac{u^2}{2}\right) \tag{6.1}$$

The enthalpy increase ΔH can be replaced through the integral form of Equation (1.19), thus changing (6.1) to

$$q = \int_{T_1}^{T_2} TdS + \int_{P_1}^{P_2} \frac{dP}{\rho} + \frac{u_2^2 - u_1^2}{2} \tag{6.2}$$

The irreversibilities or "lost work" between points 1 and 2 can be defined as

$$lw_{12} \equiv \int_{T_1}^{T_2} TdS - q \tag{6.3}$$

where

lw = irreversibilities per unit mass of fluid, J/kg or Nm/kg or m^2/s^2

Then from Equations (6.2) and (6.3)

$$lw_{12} = \int_{P_2}^{P_1} \frac{dP}{\rho} + \frac{u_1^2 - u_2^2}{2} \tag{6.4}$$

Equation (6.4) is the *mechanical energy balance* for free, horizontal flow. For an incompressible fluid Equation (6.4) reduces to

$$lw_{12} = \frac{P_1 - P_2}{\rho} + \frac{u_1^2 - u_2^2}{2} \tag{6.5}$$

The irreversibilities are often expressed in terms of a dimensionless loss-coefficient defined as

$$K_{jk} \equiv \frac{2lw_j}{u_k^2} \tag{6.6}$$

where here the subscript j indicates the source of lost work and the subscript k the choice of the reference velocity.

Thus by using a mechanical energy balance the irreversibilities and hence the loss coefficient can be computed for incompressible flow from the external pressures and velocities. The downstream velocity can be related to the upstream velocity from the ratio of areas alone. The pressure change can, in turn, be determined from one or more momentum balances. Detailed derivations by this procedure are carried out for several specific cases in the following paragraphs.

It is apparent that the irreversibilities are independent of the viscosity insofar as the areas in the force-momentum balances are known and hence independent of the viscosity.

For compressible flow the integral in Equation (6.4) must be evaluated, thus requiring a relationship between P and ρ. Such calculations are suggested in Problems 7.20, 7.21 and 8.18.

Irreversibilities due to sudden expansion

A sudden expansion is illustrated in Fig. 6.1. Point 1 is considered to be sufficiently upstream so as to be unaffected by the expansion. Point 2 is located just inside the larger tube. Point 3 is located sufficiently downstream so that the flow is again fully developed.

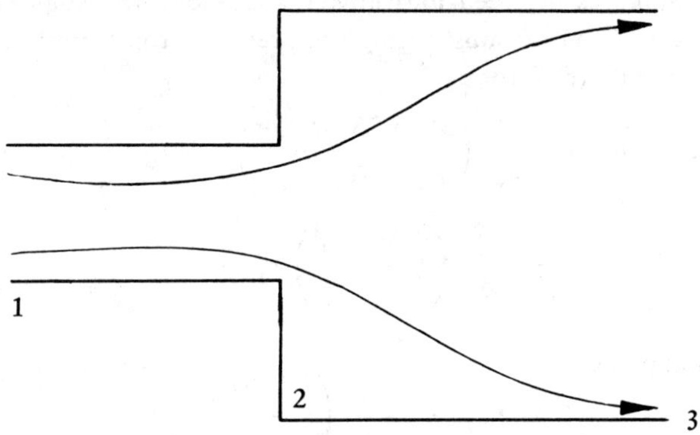

Fig. 6.1. Sudden expansion

The pressure at location 2 will be assumed to be the same outside as well as inside the jet stream and will further be assumed to differ negligibly from that at location 1. Similarly the momentum of the jet of fluid at location 2 will be assumed to differ negligibly from that at location 1. The viscous drag of the wall on the fluid will be assumed to be negligible both upstream and downstream from the contraction. Then the force-momentum balance between locations 2 and 3 can be written entirely in terms of conditions at 1 and 3 as follows:

$$P_1 A_3 + (u_1 A_1 \rho) u_1 = P_3 A_3 + (u_1 A_1 \rho) u_3 \tag{6.7}$$

or

$$P_3 - P_1 = (A_1/A_3)\rho u_1 (u_1 - u_3) \tag{6.7a}$$

The mass balance between 1 and 3 can be written as

$$u_1 A_1 \rho \ A_1 \rho = u_3 A_3 \rho \tag{6.8}$$

Eliminating u_3 between (6.7a) and (6.8) gives

$$P_3 - P_1 = \rho u_1^2 \left(\frac{A_1}{A_3}\right)\left(1 - \frac{A_1}{A_3}\right) \tag{6.9}$$

For round pipes Equation (6.9) can be written as

$$P_3 - P_1 = \rho u_1^2 \left(\frac{D_1}{D_3}\right)^2 \left[1 - \left(\frac{D_1}{D_3}\right)^2\right] \tag{6.10}$$

The irreversibilities between locations 1 and 3 due to expansion are obtained from Equation (6.5) by substituting the pressure drop from Equation (6.9) and u_3 from Equation (6.8) for u_2:

$$\begin{aligned} lw_E = lw_{13} &= u_1^2 \left(\frac{A_1}{A_3}\right)\left(\frac{A_1}{A_3} - 1\right) + \frac{u_1^2}{2}\left(1 - \left(\frac{A_1}{A_3}\right)^2\right) \\ &= \frac{u_1^2}{2}\left(1 - \frac{A_1}{A_3}\right)^2 \end{aligned} \tag{6.11}$$

or for round pipes

$$lw_E = \frac{u_1^2}{2}\left(1 - \left(\frac{D_1}{D_3}\right)^2\right)^2 \tag{6.12}$$

Equation (6.11) is known as the Borda-Carnot equation. Equation (6.12) has been shown by Schutt [1] and Kays [2] to be in good agreement with experimental data for $D_1 u_1 \rho/\mu > 10{,}000$. For lower Reynolds numbers lw_E is somewhat less owing to the greater variation in the local velocity across the channel.

The loss-coefficient for expansion based on u_1 is correspondingly

$$K_{E1} \equiv \frac{2lw_E}{u_1^2} = \left(1 - \frac{A_1}{A_3}\right)^2 \tag{6.13}$$

Irreversibilities due to sudden contraction

The process of over-contraction and reexpansion following a sudden decrease in cross-sectional area is illustrated in Fig. 6.2. Here Point 1 represents upstream conditions, Point 2 is just inside the smaller channel, Point 3 is the location of the *vena contracta* and Point 4 represents downstream conditions.

Equation (2.5) with $(A_1/A_2)^2$ substituted for $(D_1/D_o)^4$ can be assumed to approximate the contraction from location 1 to location 3. Rearrangement and elimination of the subscript on ρ gives

$$P_1 - P_3 = \frac{\rho u_1^2}{2C_o^2}\left[\left(\frac{A_1}{A_2}\right)^2 - 1\right] \tag{6.14}$$

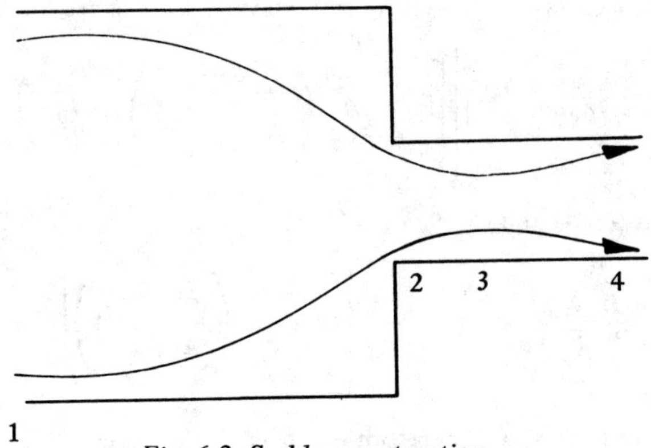

Fig. 6.2. Sudden contraction

A force-momentum balance for the expansion from location 3 to 4 can be written by the same procedure as in the previous section. The pressure at location 3 is assumed to be uniform both inside and outside the jet and the drag of the wall on the fluid is assumed to be negligible. Then

$$P_3 A_4 + (\rho u_1 A_1) u_3 = P_4 A_4 + (\rho u_1 A_1) u_4 \tag{6.15}$$

or

$$P_4 - P_3 = \frac{\rho u_1 A_1}{A_4} (u_3 - u_4) \tag{6.15a}$$

For continuity of an incompressible fluid

$$u_1 A_1 = u_2 A_2 = u_3 A_3 = u_4 A_4 \tag{6.16}$$

Therefore

$$P_4 - P_3 = \rho u_1^2 \left(\frac{A_1}{A_4}\right)\left(\frac{A_1}{A_3} - \frac{A_1}{A_4}\right) \tag{6.17}$$

Rearranging Equation (2.6) with $(A_1/A_2)^2$ and $(A_1/A_3)^2$ substituted for $(D_1/D_o)^4$ and $(D_1/D_2)^4$, and C_2^2 equated to 1.0 gives

$$\left(\frac{A_1}{A_3}\right) = 1 + \frac{1}{C_o^2}\left[\left(\frac{A_1}{A_2}\right)^2 - 1\right] \tag{6.18}$$

Substituting this expression for A_1/A_3 in Equation (6.17) gives

$$P_4 - P_3 = \rho u_1^2 \left(\frac{A_1}{A_4}\right)\left[\left[1 + \frac{1}{C_o^2}\left(\left(\frac{A_1}{A_2}\right)^2 - 1\right)\right]^{1/2} - \frac{A_1}{A_4}\right] \qquad (6.19)$$

From Equations (6.14) and (6.19)

$$\frac{P_1 - P_4}{\rho u_1^2} = \frac{1}{2C_o^2}\left(\left(\frac{A_1}{A_2}\right)^2 - 1\right) - \frac{A_1}{A_4}\left[\left[1 + \frac{1}{C_o^2}\left(\left(\frac{A_1}{A_2}\right)^2 - 1\right)\right]^{1/2} - \frac{A_1}{A_4}\right] \qquad (6.20)$$

But $A_2 = A_4$, and $u_2 = u_4$, therefore in terms of A_4 and u_4

$$P_1 - P_4 = \rho u_4^2\left[\frac{1}{2C_o^2}\left(1 - \left(\frac{A_4}{A_1}\right)^2\right) - \left[\left(\frac{A_4}{A_1}\right)^2 + \frac{1}{C_o^2}\left(1 - \left(\frac{A_4}{A_1}\right)^2\right)\right]^{1/2} + 1\right] \qquad (6.21)$$

The mechanical energy balance between locations 1 and 4 can be written as

$$\frac{P_4 - P_1}{\rho} + \frac{u_4^2 - u_1^2}{2} + lw_{41} = 0 \qquad (6.22)$$

Utilizing (6.21) and rearranging gives

$$lw_C = lw_{41} = \frac{P_1 - P_4}{\rho} - \frac{u_4^2}{2}\left(1 - \left(\frac{A_4}{A_1}\right)^2\right) \qquad (6.23)$$

$$= \frac{u_4^2}{2}\left[\frac{1}{C_o^2}\left(1 - \left(\frac{A_4}{A_1}\right)^2\right) - 2\left[\left(\frac{A_4}{A_1}\right)^2 + \frac{1}{C_o^2}\left(1 - \left(\frac{A_4}{A_1}\right)^2\right)\right]^{1/2} + 1 + \left(\frac{A_4}{A_1}\right)^2\right] \qquad (6.24)$$

and the loss-coefficient for a sudden contraction based on velocity u_4 is

$$K_{C4} \equiv \frac{2\,lw_C}{u_4^2} =$$

$$\frac{1}{C_o^2}\left(1-\left(\frac{A_4}{A_1}\right)^2\right) - 2\left[\left(\frac{A_4}{A_1}\right)^2 + \frac{1}{C_o^2}\left(1-\left(\frac{A_4}{A_1}\right)^2\right)\right]^{1/2} + 1 + \left(\frac{A_4}{A_1}\right)^2 \quad (6.25)$$

Martin [3] derived the exact equivalent of Equation (6.25) by a somewhat different path. He proposed the use of $C_O = 0.60$ rather than the theoretical value of 0.611. Indeed, as indicated in Fig. 6.3, this value gives better agreement with the experimental data of Kays [2] for turbulent flow, perhaps in compensation for some of the idealizations in the derivation.

Permanent loss due to an orifice

If the expansion from the *vena contracta* to the perimeter of the pipe were reversible, the pressure drop from location 1 to location 3 in Figure 2.2 would be negligible. However, the irreversibilities in this reexpansion are significant. An approximate expression for these losses can be derived by the same procedure as for a sudden contraction in channel diameter. Equations (6.14), (6.15a) and (6.16) are applicable if A_2 is replaced by A_O, A_3 by A_2, A_4 by A_3, P_3 by P_2 and P_4 by P_3, and then A_3 equated to A_1. Equations (6.17), (6.19) and (6.20) are replaced by

$$P_3 - P_2 = \rho u_1^2\left(\frac{A_1}{A_2} - 1\right) \quad (6.26)$$

$$P_3 - P_2 = \rho u_1^2\left[\left[1 + \frac{1}{C_o^2}\left(\left(\frac{A_1}{A_o}\right)^2 - 1\right)\right]^{1/2} - 1\right] \quad (6.27)$$

and

$$\frac{P_1 - P_3}{\rho u_1^2} = \frac{1}{2C_o^2}\left(\left(\frac{A_1}{A_o}\right)^2 - 1\right) - \left[1 + \frac{1}{C_o^2}\left(\left(\frac{A_1}{A_o}\right)^2 - 1\right)\right]^{1/2} + 1 \quad (6.28)$$

Since in this case

$$u_3 = u_1 \quad (6.29)$$

Equation (6.5) gives

$$lw_O = \frac{P_1 - P_3}{\rho} \tag{6.30}$$

and the loss-coefficient for the orifice based on the velocity in the pipe is

$$K_{O1} \equiv \frac{2lw_O}{u_1^2} = \frac{1}{C_O^2}\left(\left(\frac{A_1}{A_O}\right)^2 - 1\right) - 2\left[1 + \frac{1}{C_O^2}\left(\left(\frac{A_1}{A_O}\right)^2 - 1\right)\right]^{1/2} + 2 \tag{6.31}$$

The loss over an orifice is sometimes expressed as the ratio of the overall pressure drop to the measured pressure drop. Thus, in this case as

$$\frac{P_1 - P_2}{P_1 - P_2} = 1 - \frac{2C_O^2\left[\left[1 + \frac{1}{C_O^2}\left(\left(\frac{A_1}{A_O}\right)^2 - 1\right)\right]^{1/2} - 1\right]}{\left(\frac{A_1}{A_O}\right)^2 - 1} \tag{6.32}$$

Equation (6.32) with $C_O = 0.611$ appears to be in good agreement with the recommended curve in the ASME Power Test Code [4].

Problems

6.1. Calculate the net pressure drop, lost work and percentage loss for Problem 2.3b.

6.2. Estimate the pressure change, lost work and loss coefficient for a 1.5 m/s stream of water
a) in expansion from a 25 mm pipe to a 50 mm pipe.
b) in contraction back to the 25 mm pipe.

6.3. Integration of Equation (1.1) for constant density and horizontal flow gives

$$P_1 - P_2 = \rho(u_2^2 - u_1^2)/2$$

which is not equivalent to Equation (6.7a). Explain.

6.4. Why does laminar flow decrease the expansion-loss coefficient? Does it decrease the compression-loss coefficient? Explain.

6.5. Bird et al. [5] give the expressions below for the loss-coefficient which they attribute to Kramers. Compare with the expressions derived herein for A_1/A_4 and $A_O/A_1 = 0, 0.25, 0.5, 0.75$ and 1.0.

$$K_{C4} = 0.45\left(1 - \frac{A_1}{A_4}\right) \tag{6.33}$$

$$K_{o1} = 2.7\left(1 - \frac{A_o}{A_1}\right)\left(1 - \left(\frac{A_o}{A_1}\right)^2\right)\left(\frac{A_1}{A_o}\right)^2 \tag{6.34}$$

Explain how Equations (6.33) and (6.34) may have been developed.

6.6. *Perry's Handbook* [6], p. 5-11, recommends $1 - (D_O/D_1)^2$ for the fractional loss of the pressure drop across an orifice. Compare this expression and Equation (6.32) with the curve given by the ASME Power Test Code [4]. Values read from this curve are

$\frac{D_O}{D_1}$	$\frac{P_1 - P_3}{P_1 - P_2}$
0.2	0.94
0.3	0.883
0.4	0.815
0.5	0.73
0.6	0.63
0.7	0.52
0.8	0.40

6.7. *Perry's Handbook* [6], p. 5-30, gives the following values for a sudden contraction in turbulent flow. Compare with Equation (6.25).

A_4/A_1	K_{C1}
0	0.5
0.2	0.45
0.4	0.36
0.6	0.21
0.8	0.07
1.0	0

6.8. Pressure losses due to expansion and compression are sometimes expressed in terms of equivalent lengths of straight pipe. Reexpress Equations (6.9)

or (6.11), (6.21) and (6.24) in equivalent lengths.

6.9. Derive expressions for **lw** and **K** for a round Borda mouthpiece as shown below. Lamb [7] derived a value of 0.5 for the entrance coefficient of such a round or planar entrance.

Fig. 6.3. Borda mouthpiece

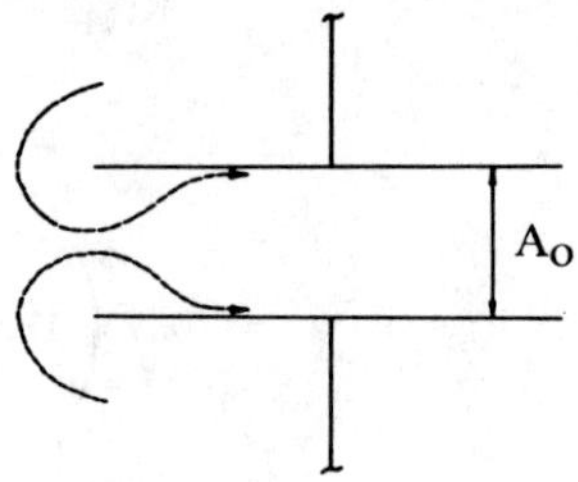

6.10. Equation (6.32), for the permanent pressure loss due to an orifice, does not contain the viscosity. Explain briefly.

6.11. None of the above expressions developed for the irreversibilities associated with sudden expansions and contractions include the viscosity explicitly. Explain briefly.

6.12. Water is flowing at 8.16 m^3/day through a 100 mm pipeline. The flow is measured by means of a removable sharp-edged orifice, 38 mm in diameter. Compute the additional yearly pumping cost resulting from the continuous use of the orifice. The pump which operates 24 hours per day for 300 days per year, has a combined fluid, mechanical and electrical efficiency of 55%. The marginal cost of electricity is 11.6 cents per KWH.

6.13. Show that the loss coefficient for contraction from a tank to a pipe approaches $4/\pi^2$ as Re increases.

6.14. Derive an expression for the loss-coefficient for the flow of Figure 2.5 in terms of z_1 and d only.

6.15. Derive an expression for the loss-coefficient for the flow of Figure 2.6 in terms of z_1, z_2 and d only.

6.16. Derive an expression for the loss coefficient for the sudden expansion of an open-channel flow.

6.17. Derive an expression for the loss-coefficient for the sudden contraction of an open-channel flow.

6.18. A sluice gate is lowered into a stream of water flowing through a wide channel. Derive an expression for the resulting permanent change in depth and velocity, if any.

References

1. H. C. Schutt, "Losses of Pressure Head Due to Sudden Enlargement of a Flow Cross-Section," *Trans. ASME* **51**, HYD-51-10, 83-87 (1929).
2. W. M. Kays, "Loss Coefficients for Abrupt Changes in Flow Cross Section with Low Reynolds Number Flow in Single and Multiple-Tube Systems," *Trans. ASME* **72**, 1067-1074 (1950).
3. J. J. Martin, "Expansion and Contraction Losses in Fluid Flow," *Chem. Eng. Educ.* **8**, No. 3, 138-140, 148 (Summer 1974).
4. *ASME Power Test Code 19, Supplement on Instruments and Apparatus,* Part 5, *Measurement of Quantity of Materials,* Chapter 4, Flow Measurement by Means of Thin Plate Orifices, Flow Nozzles and Venturi Tubes, ASME, New York (1959).
5. R. B. Bird, W. E. Stewart and E. N. Lightfoot, *Transport Phenomena*, Wiley, New York (1960), p. 217.
6. J. H. Perry, Ed., *Chemical Engineers' Handbook*, 4th Ed., McGraw-Hill, New York (1963).
7. H. Lamb, *Hydrodynamics*, 1st American Edition, Dover, New York (1945), p. 96f.

Chapter 7

Shock waves

Explosions, electrical discharges and the movement of bodies at supersonic speed create compression waves. Under some conditions these waves coalesce into a single strong wave propagating at supersonic speed. Such a wave is called a shock wave and exhibits a near discontinuity in pressure, density and temperature. The acoustic waves described in Chapter 3 represent the limiting case of shock waves as the pressure disturbance approaches zero. The behavior associated with finite pressure changes is considered in this chapter. Such waves are important because of their potential for damage but also because of their use as a device for the production of high temperature in the laboratory.

Model of an idealized shock wave

The wave is herein postulated to be a finite discontinuity in ρ, P, T and u, i.e., eddy and molecular transport are neglected. Ideal gas behavior is also assumed. The symbols used to describe such a wave moving horizontally through a stagnant gas are shown in Fig. 7.1.

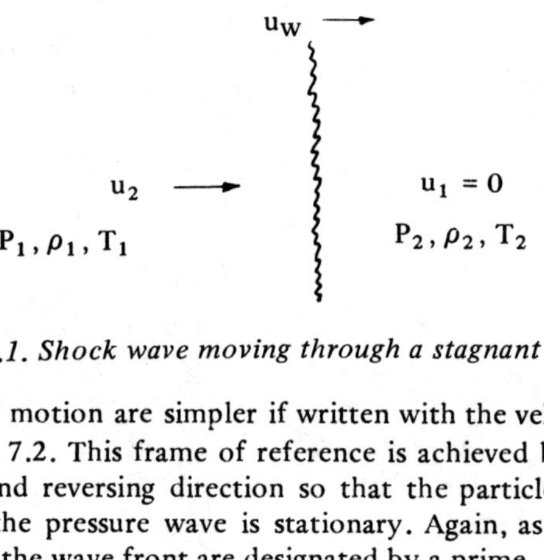

Fig. 7.1. Shock wave moving through a stagnant gas

The equations of motion are simpler if written with the velocities referred to the wave as in Fig. 7.2. This frame of reference is achieved by subtracting the wave velocity u_W and reversing direction so that the particle velocities have a positive value and the pressure wave is stationary. Again, as in Chapter 3, the velocities relative to the wave front are designated by a prime.

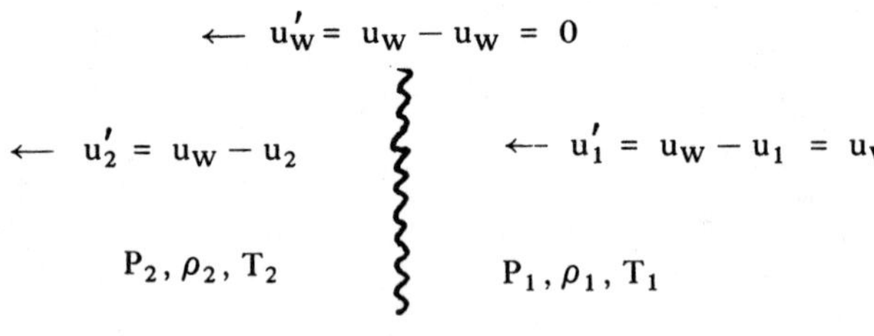

Fig. 7.2. Stationary frame of reference for a shock wave

Mass, force-momentum and energy balances can be written in terms of the variables in Figure 7.2 as follows:

$$u_1'\rho_1 = u_2'\rho_2 \tag{7.1}$$

$$P_1 + u_1'^2\rho_1 = P_2 + u_2'^2\rho_2 \tag{7.2}$$

$$H_1 + \frac{u_1'^2}{2} = H_2 + \frac{u_2'^2}{2} \tag{7.3}$$

For an ideal gas, no reaction and constant cp

$$H_2 - H_1 = cp(T_2 - T_1) \tag{7.4}$$

$$= \frac{cpm}{R}\left(\frac{P_2}{\rho_2} - \frac{P_1}{\rho_1}\right) = \frac{\gamma}{\gamma - 1}\left(\frac{P_2}{\rho_2} - \frac{P_1}{\rho_1}\right) \tag{7.5}$$

Shock wave relationships

Equations (7.1) and (7.2) can be combined to give

$$u_1'\rho_1 = u_2'\rho_2 = \left(\frac{P_2 - P_1}{\dfrac{1}{\rho_1} - \dfrac{1}{\rho_2}}\right)^{1/2} \tag{7.6}$$

Eliminating u_1', u_2', H_1 and H_2 in Equations (7.3) through (7.5) and (7.6) yields the following relationship between P_2 and ρ_2:

$$\frac{P_2}{P_1} = \frac{\beta\dfrac{\rho_2}{\rho_1} - 1}{\beta - \dfrac{\rho_2}{\rho_1}} \tag{7.7}$$

where $\beta = (\gamma + 1)/(\gamma - 1)$. Equation (7.7) is known as the *Hugoniot* or *Rankine-Hugoniot* equation* and is analogous to Equation (1.37) for an isentropic compression or expansion.

**Named after the French physicist H. Hugoniot [1] who completed the elementary theory of shock waves begun by Rankine [2] and others.*

It follows that

$$\frac{u_1'}{u_2'} = \frac{\rho_2}{\rho_1} = \frac{1 + \beta \dfrac{P_2}{P_1}}{\beta + \dfrac{P_2}{P_1}} \tag{7.8}$$

and

$$\frac{T_2}{T_1} = \frac{\beta + \dfrac{P_2}{P_1}}{\beta + \dfrac{P_1}{P_2}} = \frac{\beta - \dfrac{\rho_1}{\rho_2}}{\beta - \dfrac{\rho_2}{\rho_1}} = \frac{\beta - \dfrac{u_2'}{u_1'}}{\beta - \dfrac{u_1'}{u_2'}} \tag{7.9}$$

Alternatively

$$\frac{P_2}{P_1} = \frac{\beta \dfrac{u_1'}{u_2'} - 1}{\beta - \dfrac{u_1'}{u_2'}} \tag{7.10}$$

Thus all of the ratios are known if any one is.

It is also useful to express these ratios in terms of Mach numbers, ahead of and behind the wave, defined as follows:

$$M_1'^2 = \frac{u_1'^2 \rho_1}{\gamma P_1} = \frac{u_W^2 \rho_1}{\gamma P_1} \tag{7.11}$$

and

$$M_2'^2 = \frac{u_2'^2 \rho_2}{\gamma P_2} \tag{7.12}$$

It should be emphasized that M_1' and M_2' are defined in terms of the *stationary* wave and in terms of the properties of the uncompressed and compressed gas, respectively.

Substituting for u_1', in Equation (7.11) from the right-hand side of (7.6) and then for ρ_2/ρ_1 from the right-hand side of (7.8) and simplifying gives

$$M_1'^2 = \frac{\gamma-1}{2\gamma}\left(1 + \beta\frac{P_2}{P_1}\right) \tag{7.13}$$

Similarly,

$$M_2'^2 = \frac{\gamma-1}{2\gamma}\left(1 + \beta\frac{P_1}{P_2}\right) \tag{7.14}$$

The Mach numbers can thus be calculated from the pressure ratio or, in view of Equations (7.8) and (7.9), from the temperature, density or velocity ratios across the wave. Conversely, all of these ratios can be expressed in terms of M_1' as follows:

$$\frac{P_2}{P_1} = 1 + \frac{2\gamma}{\gamma+1}\left(M_1'^2 - 1\right) \tag{7.15}$$

$$\frac{u_2'}{u_1'} = \frac{\rho_1}{\rho_2} = 1 - \frac{2}{\gamma+1}\left(1 - \frac{1}{M_1'^2}\right) \tag{7.16}$$

$$\frac{T_2}{T_1} = \left(1 + \frac{2\gamma}{\gamma+1}\left(M_1'^2 - 1\right)\right)\left(1 - \frac{2}{\gamma+1}\left(1 - \frac{1}{M_1'^2}\right)\right) \tag{7.17}$$

and

$$M_2'^2 = \frac{1 + \dfrac{\gamma-1}{\gamma+1}\left(M_1'^2 - 1\right)}{1 + \dfrac{2\gamma}{\gamma+1}\left(M_1'^2 - 1\right)} \tag{7.18}$$

The entropy change across the wave can be related to M_1' as follows. From Equation (1.19)

$$dS = \frac{dH}{T} - \frac{dP}{\rho T} = \frac{c_P dT}{T} - \frac{R}{m}\frac{dP}{P} \tag{7.19}$$

Integrating for constant c_P and substituting for R/m gives

$$S_2 - S_1 = c_P \ln\left\{\frac{T_2}{T_1}\right\} - (c_P - c_V)\ln\left\{\frac{P_2}{P_1}\right\} = c_P \ln\left\{\frac{\rho_1}{\rho_2}\right\} + c_V \ln\left\{\frac{P_2}{P_1}\right\} \tag{7.20}$$

or in terms of M_1'

$$\frac{S_2 - S_1}{c_P} = \ln\left\{1 - \frac{2}{\gamma+1}\left(1 - \frac{1}{M_1'^2}\right)\right\} + \frac{1}{\gamma}\ln\left\{1 + \frac{2\gamma}{\gamma+1}(M_1'^2 - 1)\right\} \tag{7.21}$$

Fig. 7.3 is a plot of Equation (7.21) for $\gamma = 1.4$.

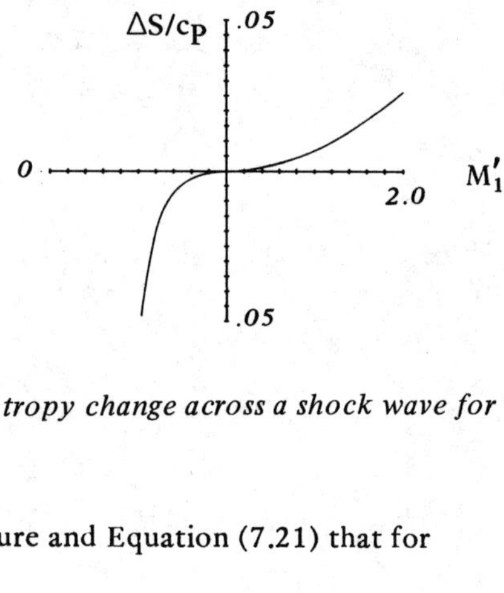

Fig. 7.3. Entropy change across a shock wave for $\gamma = 1.4$

It is apparent from this figure and Equation (7.21) that for

$M_1' > 1$ $\qquad$ $S_2 - S_1 > 0$

$M_1' = 1$ $\qquad$ $S_2 - S_1 = 0$

and

$M_1' < 1$ $\qquad$ $S_2 - S_1 < 0$

Therefore M_1' must be greater than unity. Since $u_W = u_1'$, a shock wave is necessarily supersonic with respect to the stagnant gas. It follows from Equations

(7.15) - (7.18) that $P_2 > \rho_1$, $\rho_2 > 1$, $T_2 > T_1$, $u_2' < u_1'$ and $M_2' > 1$. A shock wave is thus necessarily a compression wave. A shock expansion is not possible.

The behavior of shock waves may be illustrated graphically in a so-called *Hugoniot* plot of P versus $1/\rho$ as in Fig. 7.4. The curve represents Equation (7.7) with P and ρ substituted for P_2 and ρ_2. The slope of a straight line through P_1, ρ_1 and any point P_2, ρ_2 is $(P_2 - P_1)\rho_2\rho_1/(\rho_1 - \rho_2)$ which from Equation (7.6) equals $-(u_1\rho_1)^2$. The higher the slope the greater $u_1' = u_W$. Values of $1/\rho$ greater than $1/\rho_1$ are excluded because they imply $\Delta S < 0$.

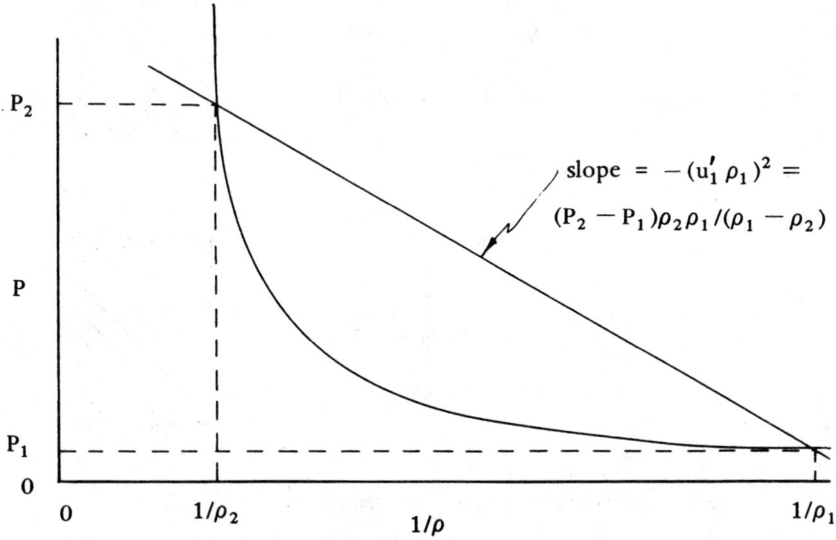

Fig. 7.4. Hugoniot plot for a shock wave

Real shock waves

Measured shock wave velocities are in excellent agreement with the values predicted by the above equations. Measured pressures and temperatures behind a shock wave are in agreement on the mean with the predicted values but some short-term, fine structure is also observed. These short-term deviations are associated with the time of relaxation of the rapidly heated gas molecules with respect to translational, rotational and vibrational motion, chemical dissociation, electronic excitation and ionization.

Reflected shock waves

Normal reflection of a shock wave off a solid body produces a further increase in pressure. Fig. 7.5 illustrates first the original shock wave before collision with the wall and then the reflected shock after collision, both in the original frame of reference relative to the surroundings as shown in Fig. 7.1. The recompressed gas in state 3 behind the reflected wave is presumed to be motionless relative to the surroundings, i.e., relative to the wall. The reflected wave encounters the previously shocked gas in state 2.

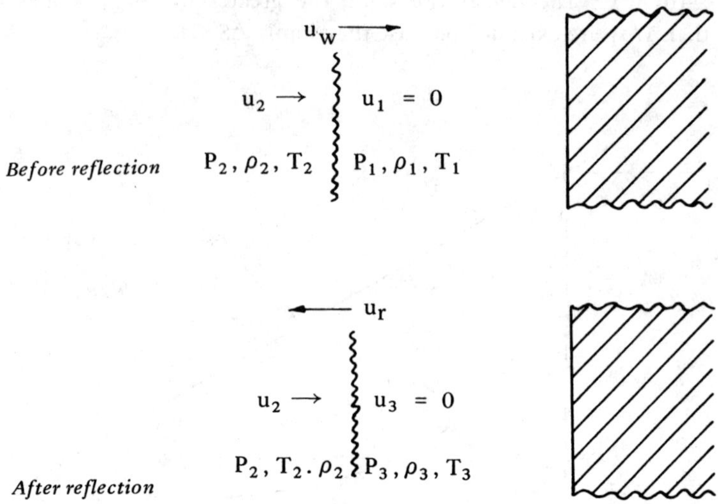

Fig. 7.5. Reflected shock wave relative to surroundings

Switching to the reflected wave as a frame of reference by subtracting the reflected wave velocity produces the relationships shown in Fig. 7.6.

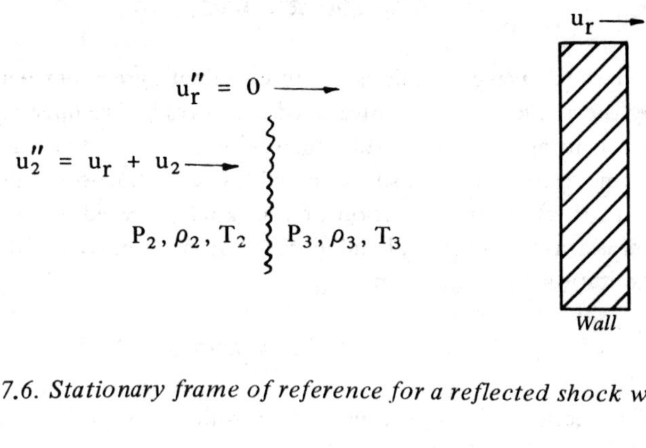

Fig. 7.6. Stationary frame of reference for a reflected shock wave

It follows by analogy to Equation (7.15) that

$$\frac{P_3}{P_2} = 1 + \frac{2\gamma}{\gamma+1}(M_r''^2 - 1) \tag{7.22}$$

where

$$M_r'' = u_2''\left(\frac{\rho_2}{\gamma P_2}\right)^{1/2} = (u_r + u_2)\left(\frac{\rho_2}{\gamma P_2}\right)^{1/2} \tag{7.23}$$

From Equations (7.1) and (7.2)

$$\frac{P_2}{P_1} = 1 + \frac{1}{P_1}(u_1'^2\rho_1 - u_2'^2\rho_2) = 1 + \frac{u_1'\rho_1}{P_1}(u_1' - u_2') \tag{7.24}$$

By analogy

$$\frac{P_3}{P_2} = 1 + \frac{u_2''\rho_2}{P_2}(u_2'' - u_3'') = 1 + \frac{u_2\rho_2}{P_2}(u_r + u_2) \tag{7.25}$$

$$= 1 + \gamma M_r'' M_2 \tag{7.26}$$

where

$$M_2 = u_2\left(\frac{\rho_2}{\gamma P_2}\right)^{1/2} = (u_1' - u_2')\left(\frac{\rho_2}{\gamma P_2}\right)^{1/2} = M_1'\left(\frac{T_1}{T_2}\right)^{1/2} - M_2' \tag{7.27}$$

Eliminating M_r'' between Equations (7.22) and (7.26) and solving the resulting quadratic gives

$$\frac{P_3}{P_2} = 1 + \frac{\gamma(\gamma+1)(M_2)^2}{4}\left[1 + \left[1 + \left(\frac{4}{M_2(\gamma+1)}\right)^2\right]^{1/2}\right] \tag{7.28}$$

Then from Equation (7.26)

$$M_r'' = \frac{(\gamma+1)M_2}{4}\left[1 + \left[1 + \left(\frac{4}{M_2(\gamma+1)}\right)^2\right]^{1/2}\right] \tag{7.29}$$

By analogy to Equations (7.8) and (7.16)

$$\frac{u_3''}{u_2''} = \frac{\rho_2}{\rho_3} = \frac{\beta + \dfrac{P_3}{P_2}}{1 + \beta\,\dfrac{P_3}{P_2}} = 1 - \frac{2}{\gamma+1}\left(1 - \frac{1}{M_r''}\right) \tag{7.30}$$

and from Equation (7.9)

$$\frac{T_3}{T_2} = \frac{\beta + \dfrac{P_3}{P_2}}{\beta + \dfrac{P_2}{P_3}} \tag{7.31}$$

All of the conditions behind a reflected shock wave can thus be calculated from measured values of u_W or P_2/P_1. Equations (7.11) or (7.13) then give M_1', (7.15) - (7.18) give the other quantities behind the shock wave, Equation (7.27) gives M_2 and (7.28) - (7.31) give the conditions behind the reflected wave.

The behavior of real waves is of course much more complex than the above idealized description owing to transient effects, interactions with surfaces, chemical dissociation, ionization and molecular excitation. Discussions of these phenomena and reviews of the literature on shock waves are provided by Soloukhin [3], Emmons [4], Courant and Friedrichs [5], Zel'dovich and Raizer [6], and Stupochenko et al. [7].

Shock tubes

A shock tube is a device for generating shock waves in the laboratory. In its simplest form it consists of a high pressure section and a low pressure section separated by a diaphragm as shown in Fig. 7.7. When the diaphragm is ruptured a shock wave propagates from the high pressure section to the low pressure section. Taub [8] apparently first derived the following expression relating the generated shock strength, P_2/P_1, to the diaphragm pressure ratio, P_0/P_1:

$$\frac{P_0}{P_1} = \frac{P_2}{P_1}\left[1 - \left(\frac{P_2}{P_1} - 1\right)\left(\frac{\beta_1 - 1}{\beta_0 - 1}\right)\left(\frac{\gamma_1 T_1 m_0}{\gamma_0 T_0 m_1 (\beta_1 + 1)\left(1 + \beta_1 \dfrac{P_2}{P_1}\right)}\right)^{1/2}\right]^{-(\beta_0 + 1)} \tag{7.32}$$

This expression is based on the assumption that the rarefaction wave which

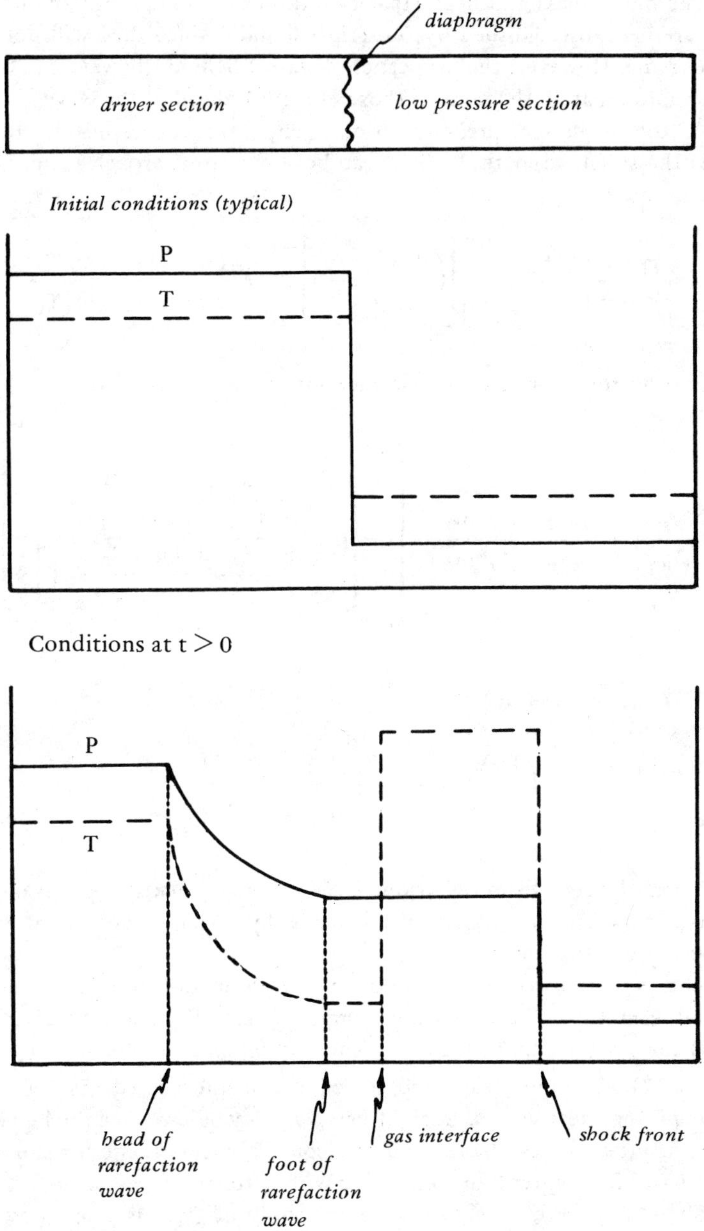

Fig. 7.7. Shock tube behavior

simultaneously propagates upstream is an isentropic expansion.

The molar mass and heat capacity ratios of the initial high and low pressure gases are here distinguished by subscripts 0 and 1 since they will not in general be the same. However, the properties of state-2 behind the wave can be assumed to be approximately the same as those of state-1 ahead of the wave.

As the diaphragm pressure ratio, P_0/P_1, approaches infinity in Equation (7.32) the shock strength, P_2/P_1, can be shown to approach a finite maximum value given by

$$\left[1+\left(\frac{\gamma_1+1}{\gamma_1-1}\right)\left(\frac{P_2}{P_1}\right)_{max}\right]\left[\left(\frac{P_2}{P_1}\right)_{max}^2-1\right]^{-1}=\frac{(\gamma_0-1)^2\gamma_1 m_0 T_1}{2(\gamma_1-1)\gamma_0 m_1 T_0} \tag{7.33}$$

The positive root of this quadratic equation is

$$\left(\frac{P_2}{P_1}\right)_{max}=1+\frac{\gamma_0(\gamma_1+1)m_1 T_1}{(\gamma_0-1)^2 m_0 T_1}\left[1+\left[1+\frac{\dfrac{2\gamma_0(\gamma_1-1)m_1 T_0}{(\gamma_0-1)^2 m_0 T_1}-1}{\left[\dfrac{\gamma_0(\gamma_1+1)m_1 T_0}{(\gamma_0-1)^2 m_0 T_1}+1\right]^2}\right]^{1/2}\right] \tag{7.34}$$

For $(T_0/m_0) > (T_1/m_1)$

$$\left(\frac{P_2}{P_1}\right)_{max}\rightarrow\frac{2\gamma_0(\gamma_1+1)m_1 T_0}{(\gamma_0-1)^2 m_0 T_1} \tag{7.35}$$

It may be seen from Equation (7.35) that the shock strength is increased by the use of a gas of low molar mass, such as hydrogen or helium, at an elevated temperature as the driver.

The potential of a shock tube for the production of high temperatures is illustrated in Table 7.1 which was computed from Equations (7.13), (7.14) and (7.17) for $\gamma = 1.4$ and $T_1 = 273$ K. The last column was computed using Equation (1.37). Decomposition and ionization of the heated gas were neglected in computing these values. Such effects, which would reduce the higher temperatures significantly, can be taken into account if chemical equilibrium is assumed, as discussed in Chapter 8 in connection with detonation waves.

Shock tubes are widely used for the study of gaseous properties and reaction rates at extreme temperatures, as discussed by Bradley [9], Soloukhin [10]. Ferri [4], Gaydon and Hurle [12], Gluckstein and Churchill [13], Resler [14] and Glass [15].

Table 7.1

Shock tube behavior

$\frac{P_2}{P_1}$	$\frac{\rho_2}{\rho_1}$	$\frac{T_2}{T_1}$	M_1'	M_2'	T_2	T_{2S}
					K	K
1.0	1.000	1.000	1.00	1.000	273	273
2.0	1.625	1.231	1.36	0.756	336	333
5.0	2.818	1.774	2.10	0.561	484	432
10.0	3.813	2.623	2.95	0.478	716	527
100.0	5.670	17.64	9.27	0.389	4814	1018
1000.0	5.965	167.64	29.28	0.379	45765	1965
∞	6.000	∞	∞	0.378	∞	∞

Problems

7.1. A shock wave is generated in a tube containing air at an initial pressure of 10 kPa and 289 K. The measured velocity of the wave is 2195 m/s. No decomposition, a constant heat capacity ratio of 1.4, and ideal gas behavior may be assumed for simplicity.

a) To what pressure and temperature will the tube be subjected if the shock wave reflects off the closed end?

b) What will be the velocity of the reflected wave relative to the tube wall?

7.2. A velocity of 1.524 km/s is measured for a planar shock wave in initially static helium at atmospheric pressure and 15.6 °C. Calculate the velocity of the gas behind the shock wave relative to the surroundings.

7.3. Does a reflected shock wave travel at supersonic or subsonic velocity relative to

a) the gas in front of the reflected wave?

b) the gas behind the reflected wave?

Substantiate your answers.

7.4. A reflected shock wave in air is observed to travel at a velocity of 416 m/s relative to the surroundings. What was the velocity of the shock wave relative to the surroundings prior to reflection? Atmospheric pressure and 289 K may be assumed.

7.5. Write mass, momentum and energy balances for a spherical shock wave in terms of velocity, pressure, density and enthalpy.

7.6. Explain briefly whether or not it is possible for two parallel planar shock waves to propagate with zero relative velocity.

7.7. A shock wave produces an increase in temperature, pressure and density. What is the source of energy for this compression and heating? Explain briefly.

7.8. Describe in detail a method of calculating the location (cross-sectional area) at which a shock wave will occur for over-expansion in a converging-diverging nozzle as a function of upstream pressure, upstream temperature, exit pressure, throat area and exit area. Isentropic expansion and isentropic compression may be assumed upstream and downstream, respectively, from the shock wave. The gas may be assumed to have a constant heat capacity ratio.

7.9. A gas having $\gamma = 1.4$ flows from a reservoir through a converging-diverging nozzle of throat area A_t. Under certain conditions, a shock wave occurs where the area is 1.2 A_t. Compute the Mach numbers on either side of the shock wave.

7.10. The shock tube reactor described by Gluckstein and Churchill [11] consists of a high pressure section containing a driver gas, a low pressure section containing a test gas and an evacuated quench chamber separated by thin diaphragms. When the first diaphragm is ruptured by compressing the driver gas, a shock wave propagates through the test gas producing an almost discontinuous increase in pressure and temperature. When the shock wave reaches and ruptures the second diaphragm the test gas flows into the quench chamber and a rarefaction wave propagates backward through the tube cooling the shocked gas and quenching the reactants. The test gas and driver gas flow down the tube behind the shock wave.This behavior is indicated in the wave diagram below.

The data below were obtained for the hydrogen-deuterium exchange reaction.

$$H_2 + D_2 \rightleftarrows 2HD \tag{7.36}$$

Because of the small heat of reaction and similar physical properties for all of the reactants and products, the temperature and pressure behind the shock wave may be considered uniform, and the shock wave, rarefaction waves and interface velocities may be considered constant.

Determine the reaction rate constant **k** assuming the second order rate expression

$$r = k\left(C_{H_2} \cdot C_{D_2} - \frac{C_{HD}^2}{4.0}\right) \tag{7.37}$$

DATA

	Test gas mol %	Product (driver-gas-free basis) mol %
H_2	14.3	9.9
D_2	18.3	13.9
A	67.4	67.4
HD	0.0	8.8
	100.0	100.0

	Initial	Reacting
T – K	293	3117
P – atm	0.20	9.00

Velocities relative to tube wall

Shock wave	2225 m/s
Rarefaction wave	685 m/s
Interface	1686 m/s

Tube length = 1.219 m

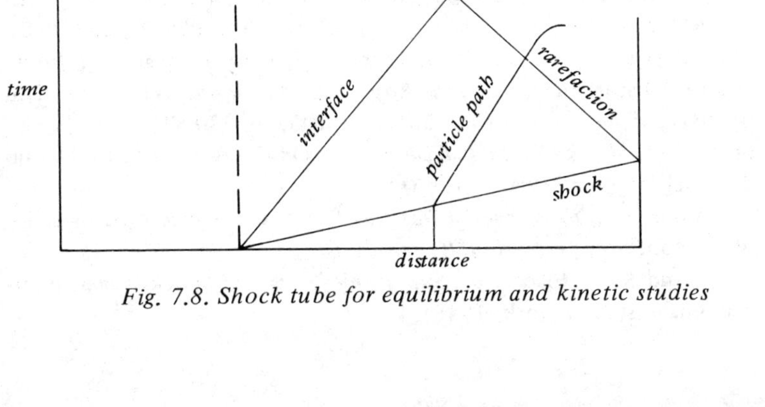

Fig. 7.8. Shock tube for equilibrium and kinetic studies

7.11. Derive expressions for the limiting pressure, density, and temperature ratios attainable with shock waves and reflected shock waves.

7.12. Calculate P_3/P_2, ρ_3/ρ_2, T_3/T_2, $M_r'' = u_r''/(\gamma RT_2/m)^{1/2}$ and T_3 if the shock wave in Table 7.1 is reflected.

7.13. Calculate the maximum attainable shock strength for

a) $m_1/m_0 = 1, T_1/T_0 = 1.$

b) $m_1/m_0 = 15, T_1/T_0 = 1.$

c) $m_1/m_0 = 1, T_1/T_0 = 2.$

d) $m_1/m_0 = 15, T_1/T_0 = 2.$

7.14. Derive expressions equivalent to Equations (7.34) and (7.35) for the maximum attainable Mach number in a shock tube.

7.15. What is the minimum possible velocity of a reflected shock wave in air? At what shock strength (P_2/P_1) does it occur?

7.16. Why does a shock wave produce a higher temperature than an isentropic expansion for the same pressure ratio?

7.17. Does the air behind a shock wave travel at supersonic or subsonic velocity relative to the surroundings? Justify your answer.

7.18. Show that the maximum possible value of M_1 in a shock tube is

$$\frac{a_0}{a_1}\left(\frac{\gamma_1+1}{\gamma_0-1}\right)$$

where the subscript 0 represents the driver pressure and the subscript 1 the unshocked gas.

7.19. Do the stagnation temperature and stagnation pressure change across a shock wave? Explain briefly.

7.20. Derive an expression for the "lost work" across a shock wave.

7.21. Derive an expression for the "lost work" across a reflected wave.

References

1. H. Hugoniot, (Memoir on the Propagation of Motion of Bodies and Particularly in Perfect Gases), *J. ecole polyt.* **57**, 3-97 (1887); **58**, 1-126 (1888).
2. W. J. M. Rankine, "On the Thermodynamic Theory of Waves of Finite Longitudinal Disturbance," *Trans. Roy. Soc. (London)* **A160**, 277-288 (1870); *Misc. Sci. Papers*, Griffin, London (1881), pp. 530-543.
3. R. I. Soloukhin, *Shock Waves and Detonations in Gases*, English Ed., Mono Book Co., Baltimore, Maryland (1966).
4. H. W. Emmons, Ed., *Fundamentals of Gas Dynamics*, Princeton University Press, Princeton, New Jersey (1958).
5. R. Courant and K. O. Friedrichs, *Supersonic Flow and Shock Waves*, Interscience Publishers, New York (1948).

6. Ya. B. Zel'dovich and Yu. P. Raizer, *Physics of Shock Waves and High-Temperature Hydrodynamic Phenomena*, English Ed., W. D. Hayes and R. F. Probstein, Eds., Academic Press, New York (1966).
7. E. V. Stupochenko, S. A. Losev and A. I. Osipov, *Relaxation in Shock Waves*, English Ed., R. S. Lee, Ed., Springer Verlag, New York (1967).
8. A. H. Taub, "Relativistic Rankine-Hugoniot Equations," *Phys. Rev.* **74**, 328-334 (1948).
9. H. N. Bradley, *Shock Waves in Chemistry and Physics*, Wiley, New York (1962).
10. R. I. Soloukhin, "Shock Waves and High-Temperature Kinetics of Gases," *American Scientist* **59**, 600-608 (1971).
11. A. Ferri, Ed., *Fundamental Data Obtained from Shock Wave Experiments*, Pergamon Press, Oxford (1961).
12. A. G. Gaydon and I. R. Hurle, *The Shock Tube in High Temperature Chemical Physics*, Chapman and Hall, London (1963).
13. M. E. Gluckstein and S. W. Churchill, "The Use of a Shock Tube for Exploratory Studies of High-Temperature Equilibrium and Reaction Rates–Hydrogen-Deuterium Exchange," *AIChE J.* **9**, 821-826 (1963).
14. Resler, E. L.,"Chemical Studies Using Gas Dynamic Wave Interactions," in *Dynamics of Fluids and Plasmas*, Academic Press, New York (1966).
15. I. I. Glass, *Shock Waves and Man*, University of Toronto Institute for Aerospace Studies, Toronto (1974).

Chapter 8

Detonation waves in gases

A detonation wave is a shock wave driven by chemical reactions. Detonations may be generated by ignition of an explosive solid, a liquid or a mixture of gas. They have greater power for destruction than isentropic explosions. Hence, their behavior is of direct interest in the processing and storage of explosive materials [1].

Idealized model of detonation wave

Figures 7.1 and 7.2 and Equations (7.1), (7.2) and (7.3) for shock waves are directly applicable to detonation waves in gases.* However, Equation (7.4) must be replaced by

$$H_2 - H_1 = c_{PU}(T_2 - T_1) + \Delta H_{R2} \qquad (8.1)$$

Here

c_{PU} = mean specific heat capacity of the unreacted gaseous mixture between T_1 and T_2, J/kg·K

ΔH_{R2} = enthalpy change due to reaction at T_2 per unit mass of mixture, J/kg or m^2/s^2.

It should be emphasized that ΔH_R is for a unit mass of the entire mixture, including any unreacted material, not the value in the handbooks for complete reaction of a mole of one of the reactants.

It is sometimes convenient to express Equation (8.1) in the alternative form

$$H_2 - H_1 = \Delta H_{R1} + c_{PR}(T_2 - T_1) \qquad (8.2)$$

where

ΔH_{R1} = enthalpy change due to reaction at T_1 per unit mass of mixture, J/kg

c_{PR} = mean heat capacity of reacted mixture between T_1 and T_2, J/kg·K.

***Hirschfelder and Curtiss [2] mention wagering with G. B. Kistiakowsky a bottle of French champagne against a case of American champagne as to whether or not transport properties can be shown to affect the behavior of a detonation wave significantly.**

Eliminating u_1' and u_2' from Equation (7.3) using (7.1) and (7.2):

$$H_2 - H_1 = \left(\frac{P_2 - P_1}{2}\right)\left(\frac{1}{\rho_1} + \frac{1}{\rho_2}\right) \qquad (8.3)$$

Substituting for $H_2 - H_1$ from Equation (8.2) and replacing ρ with Pm/RT and rearranging:

$$\frac{\Delta H_{R1}}{c_{PR}T_1} + \frac{T_2}{T_1} - 1 = \left(\frac{\gamma_R - 1}{2\gamma_R}\right)\left(\frac{P_2}{P_1} - 1\right)\left(\frac{m_R}{m_U} + \frac{T_2 P_1}{T_1 P_2}\right) \qquad (8.4)$$

Equation (8.4) is equivalent to Equation (7.7) for a shock wave. The greater complexity arises solely from the change in chemical composition.

m_R, ΔH_{R1}, c_{PR} and γ_R are functions of the final composition. Owing to the high temperature attained in the compressed gas behind a detonation wave, chemical equilibrium is ordinarily approached closely. If chemical equilibrium is assumed in the compressed gas the composition is fixed by the chemical equilibrium constants. These equilibrium constants are functions of temperature and pressure. The dependence on pressure can be avoided by expressing composition in terms of activities but that procedure introduces other complications. In order to select a sufficient set of equilibrium constants all of the chemical species which will be present must be postulated.

The equilibrium problem can be torn out of the general problem, i.e., the equilibrium composition can be calculated for chosen values of P_2 and T_2, independently of Equation (8.4). m_R, ΔH_{R1}, c_{PR} and γ_R then become known functions of T_2 and P_2 for a chosen initial composition. The problem is now reduced to one of two variables, P_2 and T_2, and one equation, (8.4). One additional relationship between P_2 and T_2 is required. For shock waves a measured value of P_2/P_1 or M_1' was found to be necessary except with a shock tube, for which case P_2/P_1 could be estimated from the diaphragm pressure ratio. For detonation waves the additional relationship is provided without experimental measurements by the postulate that S_2 is a minimum. The validity of this postulate has been confirmed by experiments for a wide variety of conditions.

Differentiating Equation (8.3) for fixed H_1, ρ_1, and P_1:

$$dH_2 = \left(\frac{1}{\rho_1} + \frac{1}{\rho_2}\right)\frac{dP_2}{2} - \left(\frac{P_2 - P_1}{2}\right)\frac{d\rho_2}{\rho_2^2} \qquad (8.5)$$

From Equation (1.19) for the same restrictions

$$dH_2 = T_2 dS_2 + dP_2/\rho_2 \tag{8.6}$$

Eliminating dH_2 between (8.5) and (8.6) and rearranging:

$$dS_2 = \left[\frac{1}{T_2}\left[\frac{1}{\rho_2} - \frac{1}{\rho_2}\right]\frac{dP_2}{2} - \left[\frac{P_2 - P_1}{2}\right]\frac{d\rho_2}{\rho_2^2}\right] \tag{8.7}$$

Setting dS_2 to zero to find the conditions for a minimum (or maximum) in entropy then gives

$$\left[\frac{dP_2}{d\rho_2}\right]_{S_2} = \left[\frac{P_2 - P_1}{\frac{1}{\rho_1} - \frac{1}{\rho_2}}\right]\frac{1}{\rho_2^2} \tag{8.8}$$

From Equation (7.6), which was obtained from (7.1) and (7.2) and hence holds for detonation waves as well as shock waves,

$$u_2'^2 = \left[\frac{P_2 - P_1}{\frac{1}{\rho_1} - \frac{1}{\rho_2}}\right]\frac{1}{\rho_2^2} = \left[\frac{dP_2}{d\rho_2}\right]_S \tag{8.9}$$

Thus the burned gas behind the detonation wave travels at acoustic velocity relative to the wave. This results, which has been confirmed by countless experiments, is known as the *Chapman-Jouguet* condition.*

A Hugoniot plot for detonation is sketched as Fig. 8.1. The curve representing Equation (8.4) does not pass through (P_1, $1/\rho_1$). A Chapman-Jouguet detonation is represented in this plot by the tangent to the curve through (P_1, $1/\rho_1$). The solution for $P_2 < P_1$ corresponds to a Chapman-Jouguet deflagration (flame). However, transport processes cannot be neglected in flames, and actual deflagrations occur to the left of the Chapman-Jouguet point.

**In recognition of the contributions of D. L. Chapman [3] and E. Jouguet [4].*

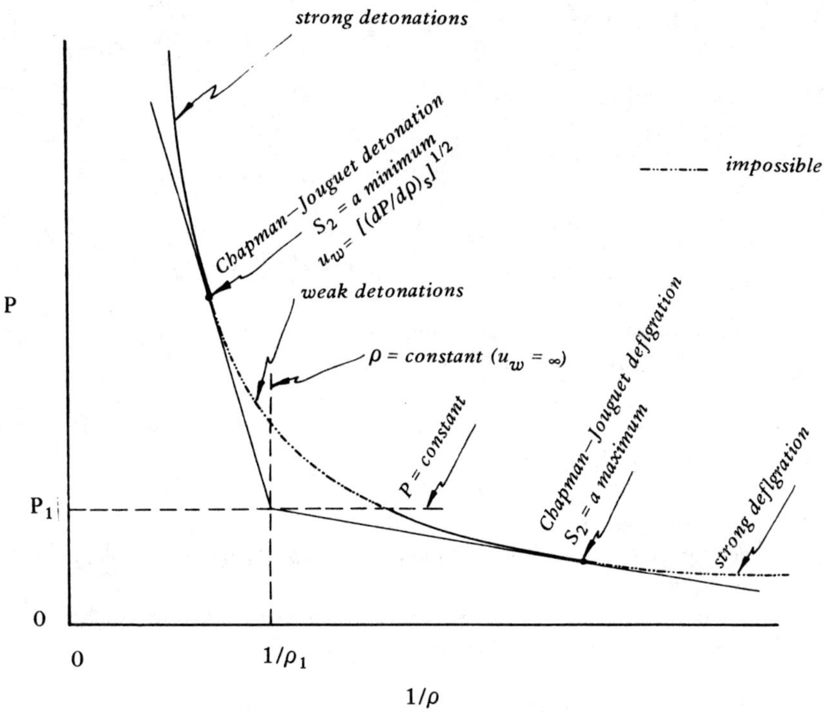

Fig. 8.1. Hugoniot plot for a detonation wave

In order to develop a calculation scheme

$$u_2' = (\gamma_R P_2/\rho_2)^{1/2} \tag{8.10}$$

is substituted in Equation (7.6) giving

$$\gamma_R P_2 \rho_2 = (P_2 - P_1)\rho_1 \rho_2/(\rho_2 - \rho_1) \tag{8.11}$$

Replacing ρ with Pm/RT and rearranging then gives a second relationship between P_2 and T_2 :

$$\frac{T_1}{T_2} = \frac{m_U}{\gamma_R m_R}\left[\gamma_R + 1 - \frac{P_1}{P_2}\right]\frac{P_1}{P_2} \tag{8.12}$$

Substituting for T_2/T_1 from Equation (8.12) in (8.4) results in

$$\frac{m_U}{m_R}\left[\frac{\Delta H_{R1}}{c_{PR}T_1}-1\right]\left[(\gamma_R+1)\frac{P_2}{P_1}-1\right]+\gamma_R\left(\frac{P_2}{P_1}\right)^2=\left[\frac{\gamma_R-1}{2}\right]\left[\frac{P_2}{P_1}-1\right]\left[\left(\frac{2\gamma_R+1}{\gamma_R}\right)\frac{P_2}{P_1}-1\right]$$

which can be simplified somewhat to (8.13)

$$\left(\frac{P_2}{P_1}\right)^2+2\left[\frac{m_U}{m_R}\left(\frac{\Delta H_{R1}}{c_{PR}T_1}-1\right)\gamma_R+\frac{\gamma_R-1}{\gamma_R+1}\left(\frac{3\gamma_R+1}{2}\right)\right]\frac{P_2}{P_1}$$
$$-\frac{m_U}{m_R}\left(\frac{\Delta H_{R1}}{c_{PR}T_1}-1\right)\frac{2\gamma_R}{\gamma_R+1}-\gamma_R\left(\frac{\gamma_R-1}{\gamma_R+1}\right)=0 \quad (8.13a)$$

The positive root of this quadratic equation gives P_2/P_1 as a function of m_R, ΔH_{R1}, c_{PR} and γ_R. A possible method of solution is:

1) For a chosen initial mixture, P_2 and T_2 calculate the equilibrium composition, then m_R, ΔH_{R1}, c_{PR} and γ_R.
2) Using these values, compute P_2/P_1 from Equation (8.13a).
3) Compute T_2/T_1 from Equation (8.12).
4) Recompute the equilibrium composition, m_R, ΔH_{R1}, c_{PR} and γ_R for the new P_2/P_1 and T_2/T_1.
5) Repeat (2), (3) and (4) until satisfactory convergence is attained.
6) From T_2/T_1 and P_2/P_1 calculate $\rho_2/\rho_1 = u_1'/u_2' = u_w/u_2'$.

The conditions resulting from reflection of a detonation wave can be computed by the same procedure used for shock waves except that the further change in composition must be taken into account. This second change in composition is however much smaller than the first and is often neglected. Obviously a computer is needed for the calculations of equilibrium, detonation, and reflection. Equivalent procedures have been utilized successfully by Lewis and von Elbe [5], Weir and Morrison [6], Moyle et al. [7], Gealer and Churchill [8], Luker et al. [9], and many others. Gealer and Churchill took deviations from the ideal gas law into account and Luker et al. took these deviations into account partially.

Calculation of equilibrium composition

As indicated, the calculation of the equilibrium composition for any chosen final pressure, temperature and initial composition can be carried out independently. Such calculations are described in most texts on thermodynamics and on process design. Hence only a brief and elementary illustration will be presented herein.

Consider the equilibrium composition at a chosen final temperature and pressure of a chosen initial mixture of O_2 and H_2. Only H_2, O_2, H_2O, H, O and OH will be assumed to be present. (The O_3 concentration can be shown to be negligible by repeating the calculations with an additional equilibrium constant). A sufficient set of equilibrium conditions and equilibrium constants for the chosen six species is

$$\text{I:}\quad \mathbf{K}_1 = \frac{C_{H_2O}^2}{C_{H_2}^2 C_{O_2}} \qquad \text{for} \qquad 2H_2 + O_2 \rightleftarrows 2H_2O$$

$$\text{II:}\quad \mathbf{K}_2 = \frac{C_{OH}^2}{C_{H_2} C_{O_2}} \qquad \text{for} \qquad H_2 + O_2 \rightleftarrows 2OH$$

$$\text{III:}\quad \mathbf{K}_3 = \frac{C_{H}^2}{C_{H_2}} \qquad \text{for} \qquad H_2 \rightleftarrows 2H$$

$$\text{IV:}\quad \mathbf{K}_4 = \frac{C_{O}^2}{C_{O_2}} \qquad \text{for} \qquad O_2 \rightleftarrows 2O$$

Conservation of hydrogen and oxygen atoms requires

$$\text{V:}\quad 2\left(\frac{C_{H_2}}{\rho}\right)_1 = \left[\frac{2C_{H_2} + 2C_{H_2O} + C_{OH} + C_H}{\rho}\right]_2$$

$$\text{VI:}\quad 2\left(\frac{C_{O_2}}{\rho}\right)_1 = \left[\frac{2C_{O_2} + C_{H_2O} + C_{OH} + C_O}{\rho}\right]_2$$

where

C_i = concentration of the i-th component, mol/m^3
K_j = equilibrium constant for j-th reaction based on concentration
ρ = density of mixture, kg/m^3.

The subscripts 1 and 2 indicate the initial and final condition.

The four equilibrium constants are presumed to be known functions of pressure and temperature, and are evaluated at P_2 and T_2. The densities and concentrations are presumed to be known functions of composition, temperature, and pressure. If density is conversely considered to be a function of concentration, temperature and pressure, the six unknown final concentrations are related by the above six equations.

More efficient methods of carrying out the equilibrium calculation, such as by minimizing the free energy [10], can of course be used for more complicated mixtures. A number of procedures for the calculation of detonation are compared by Zajak and Oppenheim [11].

Experimental confirmation

Detonation velocities are ordinarily measured in tubes of small diameter. The drag of the wall might be expected to become a significant factor as the diameter decreases. Moyle et al. [7] showed that a plot of the wave velocity versus 1/D could be extrapolated linearly to 1/D = 0 to eliminate this deviation from the above theory for unbounded waves. Fig. 8.2. is one of their plots for three $H_2 - O_2$ mixtures at 298 K and 1 atm.

Experimental detonation velocities, corrected for finite diameter, are compared with curves representing theoretical calculations in Fig. 8.3 – 8.7 for a wide range of initial conditions. The agreement is seen to be generally good.* The deviations in Fig. 8.3 at less than 20% and greater than 85% H_2 indicate the limits of detonability, which are essentially the same as the limits of flammability. These limits depend on reaction rates and hence are not predicted by the above theory which postulates equilibrium.

Values of the detonation and reflected pressure computed by Moyle et al. [7] for stoichiometric mixtures of H_2 and O_2 are compared with the average of the experimental values of Ordin [12] in Table 8.1. The agrrement is only fair but these values are very difficult to measure and the differences may be of the order of the experimental uncertainty.

Schlieren photographs of detonation waves are shown in Fig. 8.8. The wave is moving from left to right. For 52.5% H_2 small-scale turbulence can be noted behind a sharp wave front. For 30% H_2 the wave front is thicker and the scale of

---

**These values suggest that G. B.Kistiakowsky probably won or deserved to win a case of American champagne.*

Table 8.1

Comparison of computed and experimental pressures for detonation of a stoichiometric mixture of hydrogen and oxygen at 298 K and 1 atm

	Computed [7]	Experimental [12]
P_2/P_1	17.5	20.4
P_3/P_2	2.45	3.0
P_3/P_1	43.0	61.2

turbulence is greater. For 28.75% H_2 the wave has degenerated into a complicated shock pattern followed by an irregular combustion zone. In view of these photographs, it is remarkable that calculations which assume a planar wave, chemical equilibrium and negligible transfer of heat, components and momentum predict detonation velocities and pressures in such reasonable agreement with experimental values. It should be noted that the close agreement of the predicted and measured velocities does not necessarily confirm the validity of the assumptions in the model or assure that the predicted temperatures, pressures and compositions for other conditions are equally reliable.

Computed compositions and pressures

Computed equilibrium compositions for $T_1 = 298$ K, $P_1 = 1$ atm and a range of initial compositions of H_2 and O_2 are shown in Fig. 8.9. Computed compositions for a stoichiometric mixture are plotted in Fig. 8.10 for two final temperatures and a range of final pressures. It is apparent that dissociation is very significant.

Computed detonation and reflected pressures are plotted in Figs. 8.11 – 8.13, again for $H_2 - O_2$ mixtures. These pressures may be observed to be roughly proportional to the initial density, hence their severity increases with increasing initial pressure and decreasing initial temperature.

Destructive power

Luker and Liebson [13] tested rupture disks in detonation tubes and found good agreement between the calculated pressure of the reflected wave and the rated pressure of the rupture disks which just failed. Others (see, e.g., Sokolik [14], p. 352) have obtained somewhat contradictory results.

Because of the moderately high pressures generated by reflection and because of the rapid application of this pressure detonation waves can readily destroy process equipment and threaten life. Ginsburgh and Bulkley [1] and

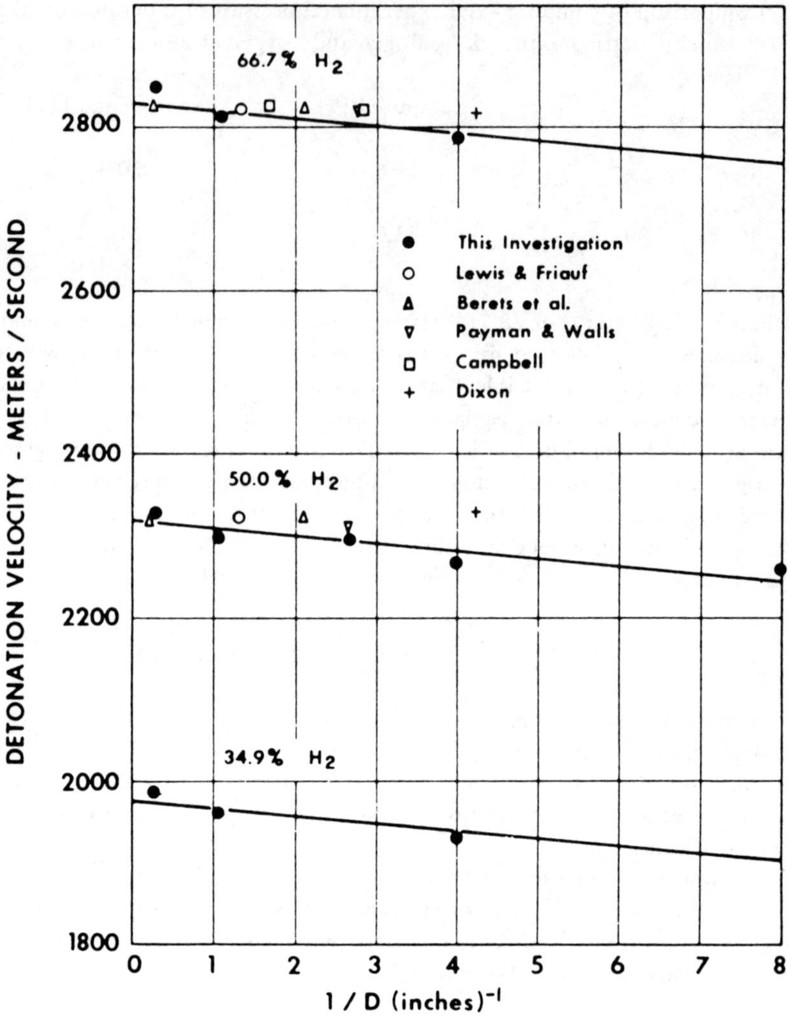

Fig. 8.2. The effect of tube diameter on the detonation velocity of $H_2 - O_2$ mixtures [7]

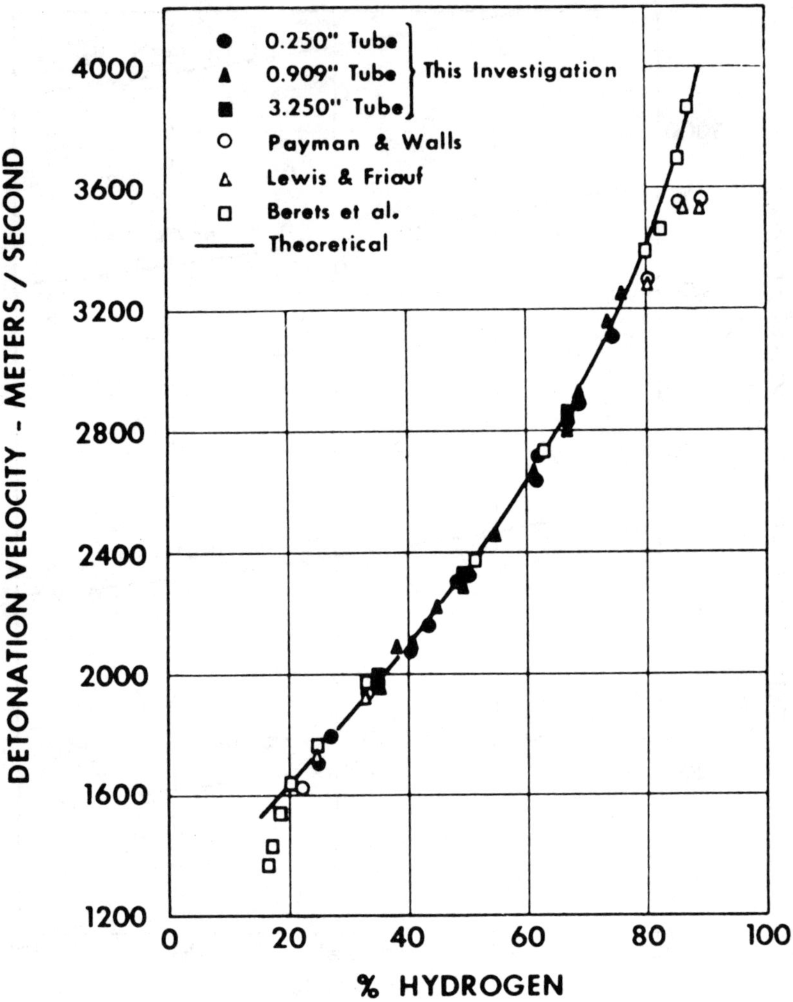

Fig. 8.3. A comparison of experimental and computed detonation velocities for $H_2 - O_2$ mixtures [7]

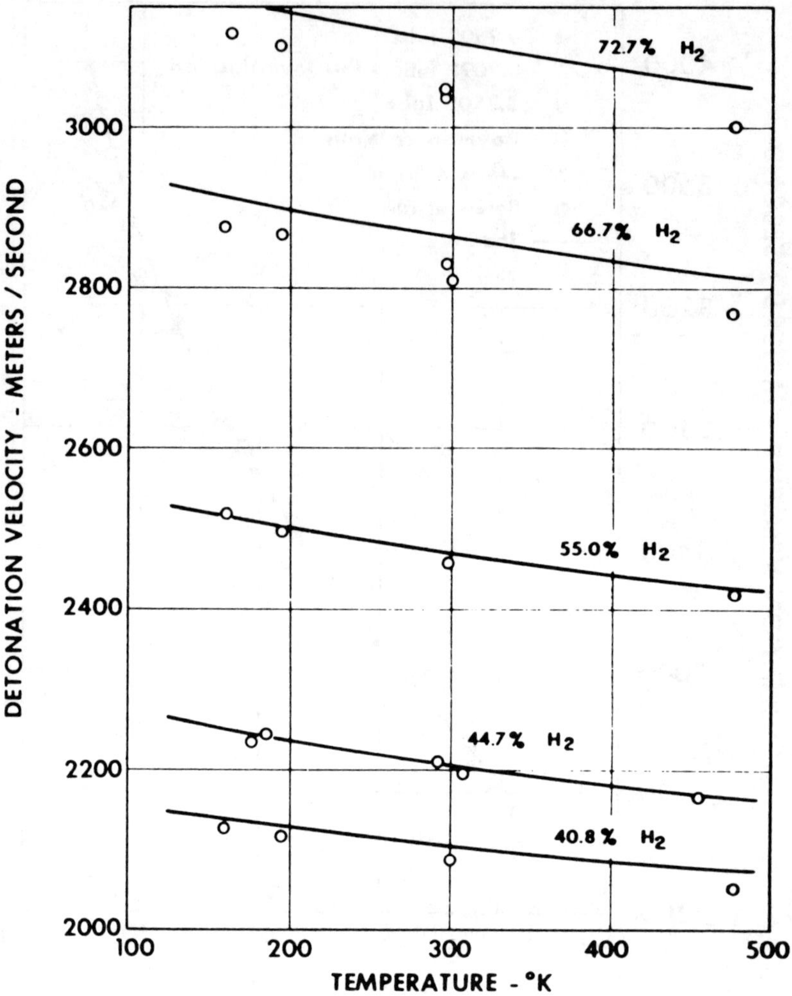

Fig. 8.4. The effect of initial temperature on the detonation velocity of H_2 – O_2 mixtures [7]

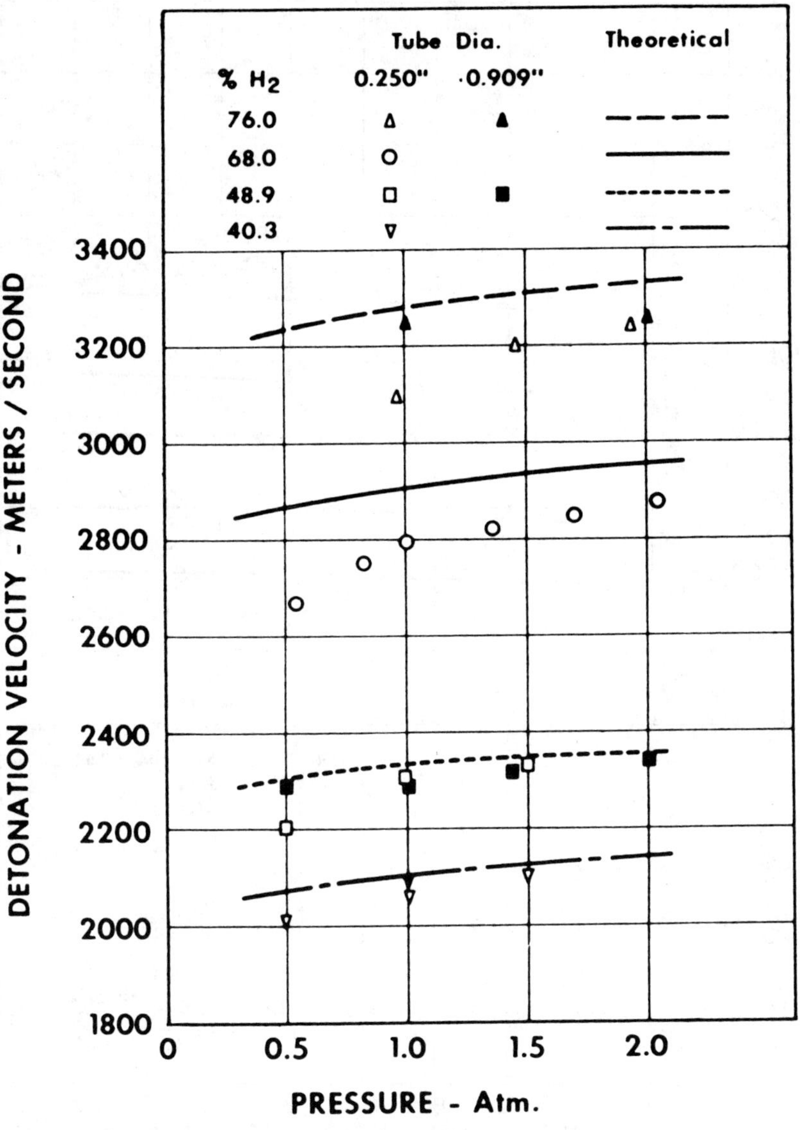

Fig. 8.5. The effect of initial pressure on the detonation velocity of $H_2 - O_2$, low range [7]

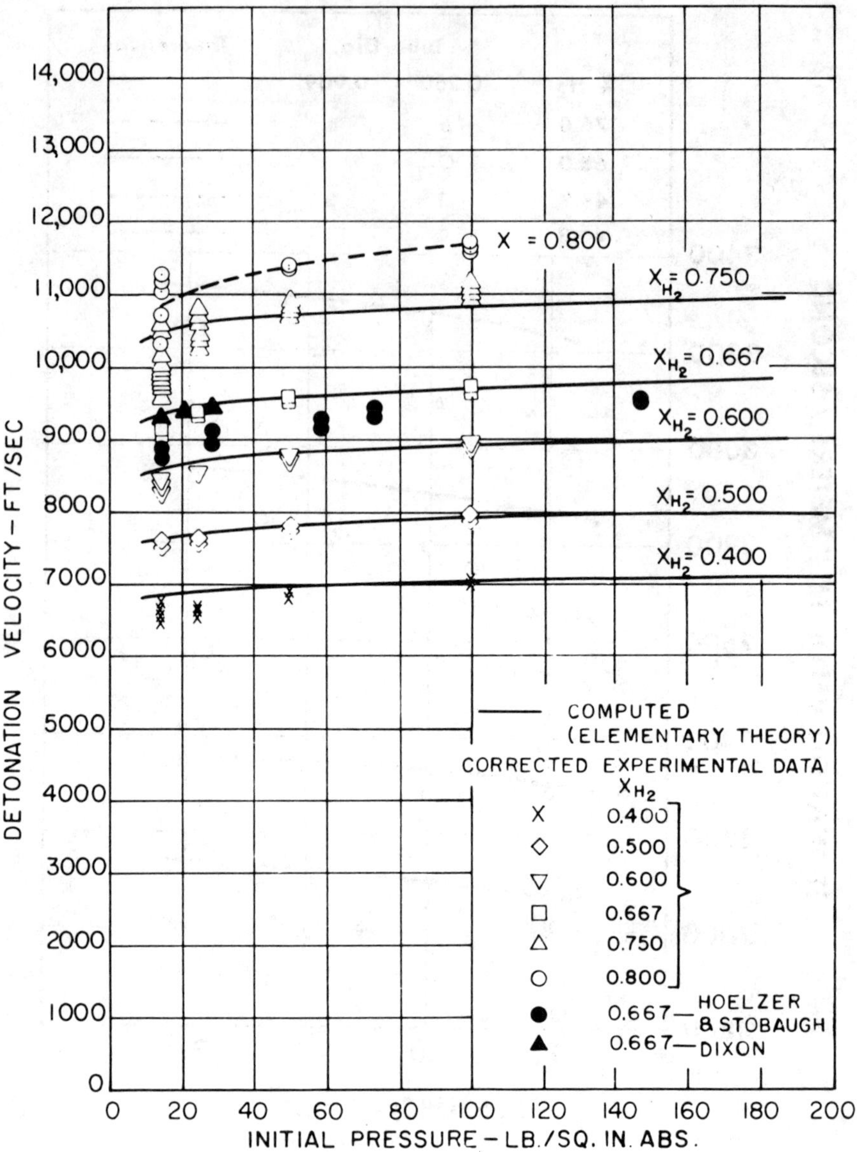

Fig. 8.6. The effect of initial pressure on the detonation velocity of $H_2 - O_2$, moderate range [8]

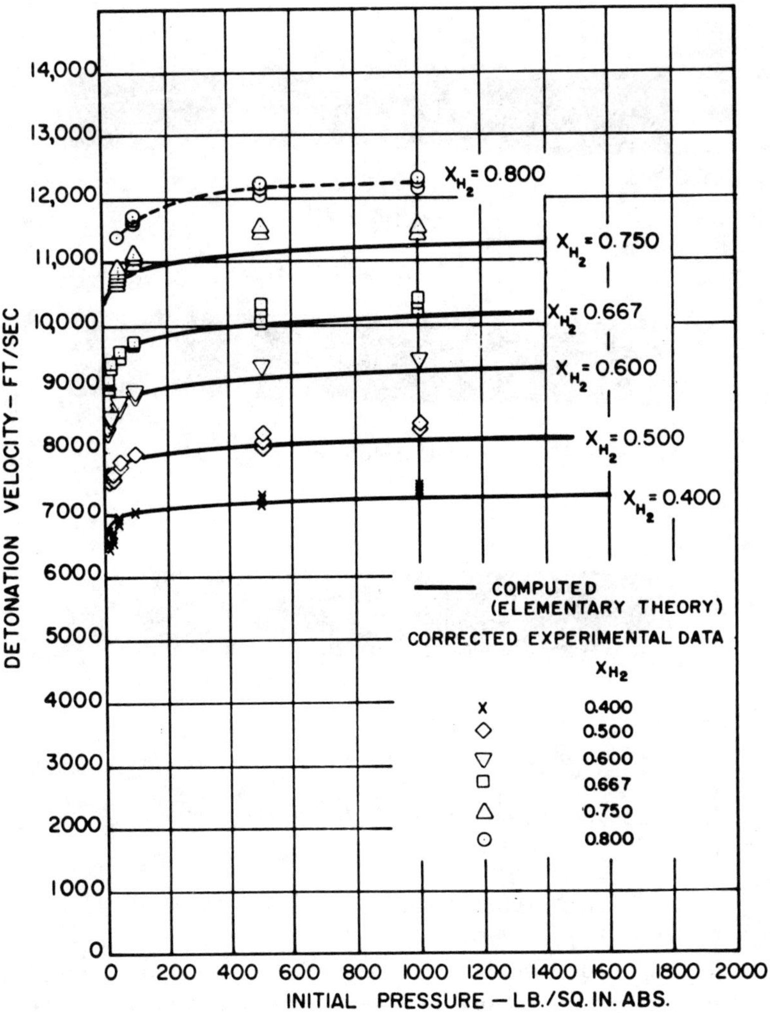

Fig. 8.7. The effect of initial pressure on the detonation velocity of $H_2 - O_2$, high range [8]

Randall et al. [15] have described such effects in industrial practice. Hence it is important to calculate the possible pressures which might be generated by detonation of a process stream or a process stream mixed with air and to take appropriate protective action. For such purposes the approximate formulation in the following section may give results of sufficient accuracy.

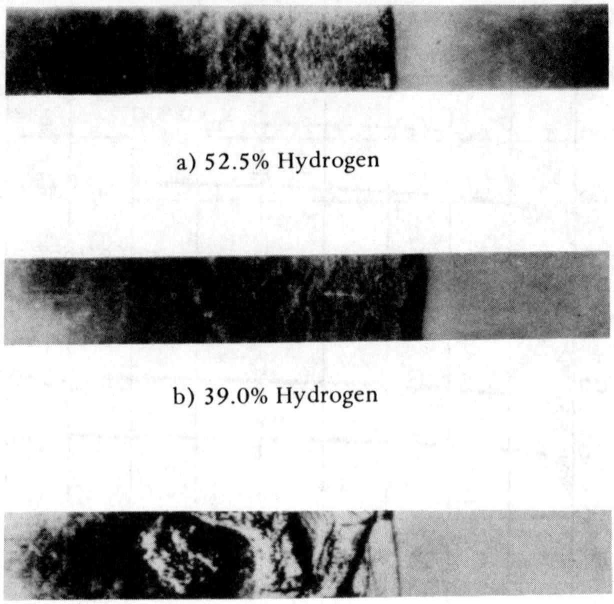

a) 52.5% Hydrogen

b) 39.0% Hydrogen

c) 28.75% Hydrogen

Fig. 8.8. Schlieren photographs of $H_2 - O_2$ detonations [7]

Approximate theory of detonation

Greatly simplified expressions for gaseous detonation waves can be derived if the change in γ and m across the wave is neglected. Equations (7.1), (7.2) and (7.3) may then be rewritten in terms of Mach numbers as follows:

$$M_1'(P_1\rho_1)^{1/2} = M_2'(P_2\rho_2)^{1/2} \tag{8.14}$$

$$P_1(1 + \gamma M_1'^2) = P_2(1 + \gamma M_2'^2) \tag{8.15}$$

and

$$c_pT_1 + \frac{\gamma P_1 M_1'^2}{2\rho_1} - \Delta H_R = c_pT_2 + \frac{\gamma P_2 M_2'^2}{2\rho_2} \tag{8.16}$$

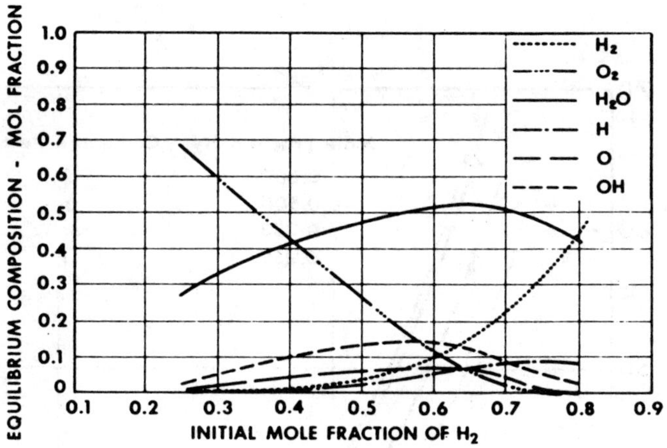

Fig. 8.9. Equilibrium compositions for detonation of $H_2 - O_2$ *mixtures* at *300 K and 1 atm [7]*

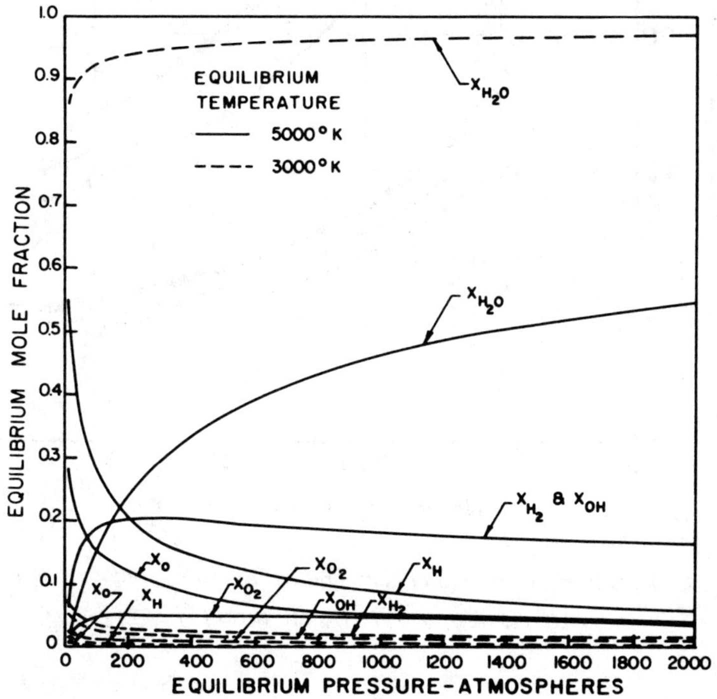

Fig. 8.10. Equilibrium compositions for a stoichiometric mixture of $H_2 - O_2$ *[8]*

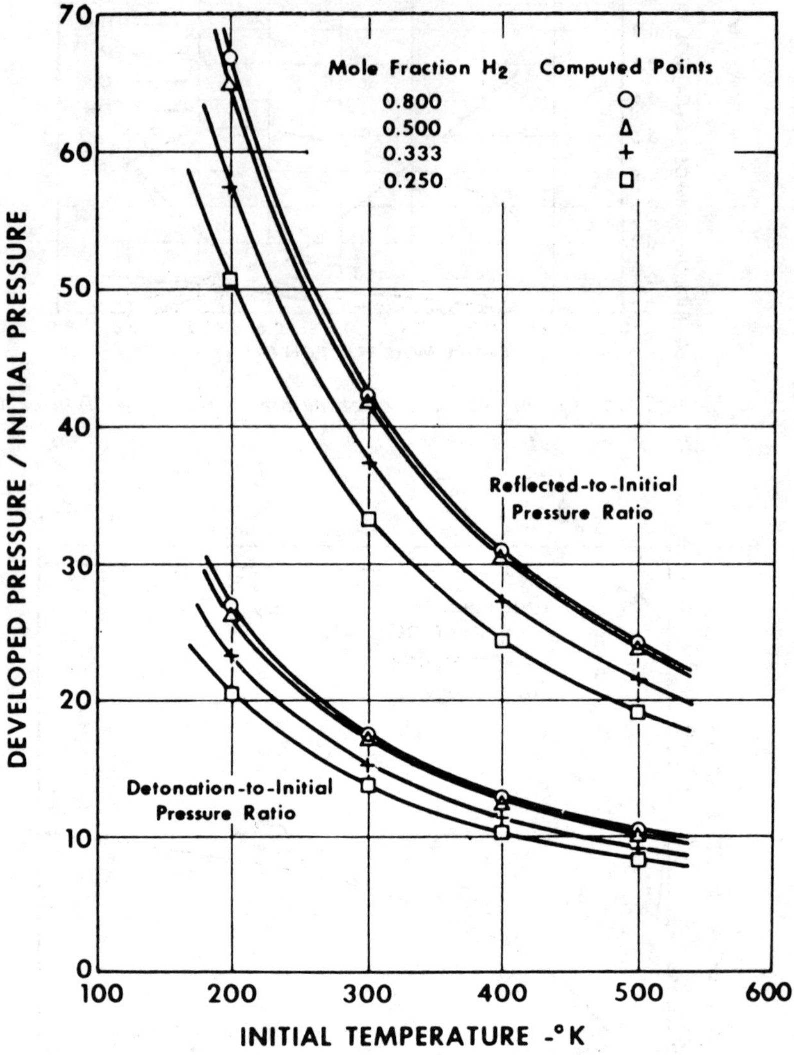

Fig. 8.11. The effect of initial temperature on computed detonation and impact pressures for $H_2 - O_2$ mixtures [7]

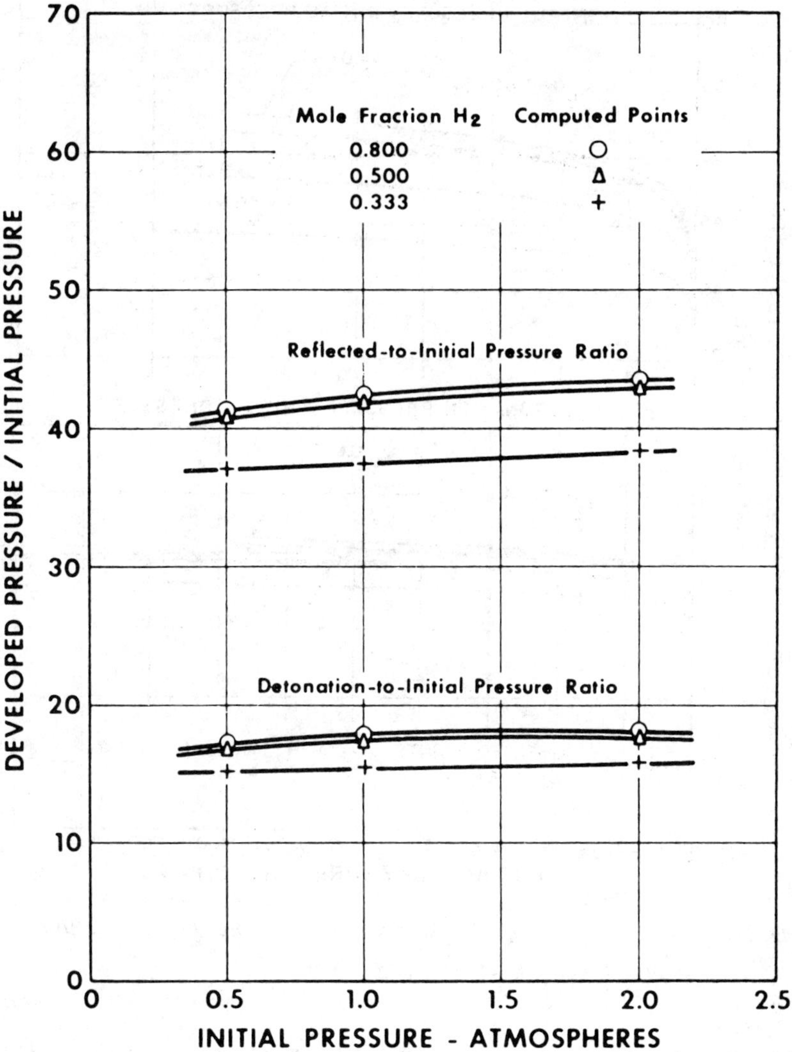

Fig. 8.12. The effect of low initial pressures on computed detonation and impact pressures for $H_2 - O_2$ mixtures [7]

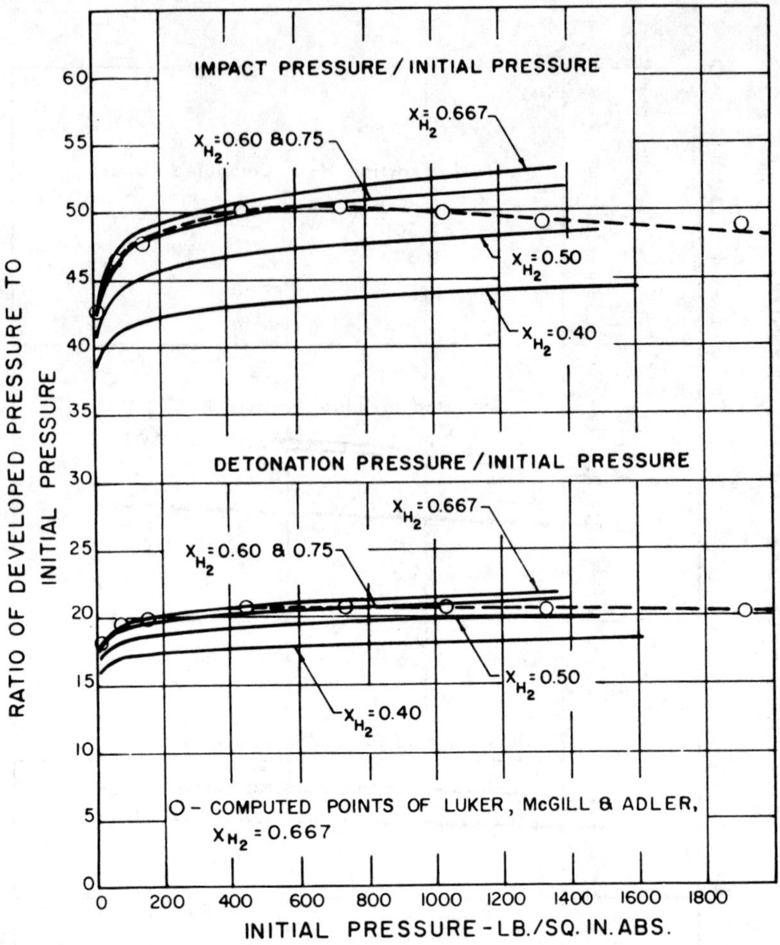

Fig. 8.13. The effect of high initial pressure on computed detonation and impact pressures for $H_2 - O_2$ mixtures [8]

where here c_p and γ are mean values for the reactants and products. This assumption of constant c_p across the wave implies that $\Delta H_{R1} = \Delta H_{R2} = \Delta H_R$.

Dividing the right and left sides of Equation (8.14) by the right and left sides, respectively, of Equation (8.15) and then squaring both sides gives

$$\frac{\rho_1}{P_1}\left(\frac{M_1'}{1+\gamma M_1^2}\right)^2 = \frac{\rho_2}{P_2}\left(\frac{M_2'}{1+\gamma M_2^2}\right)^2 \tag{8.17}$$

Multiplying the right and left sides of Equations (8.16) and (8.17) together and simplifying yields

$$\left(\frac{M_1'}{1+\gamma M_1'^2}\right)^2\left[1+\frac{\gamma-1}{2}M_1'^2-\frac{\Delta H_R}{c_pT_1}\right]=\left(\frac{M_2'}{1+\gamma M_2'^2}\right)^2\left[1+\frac{\gamma-1}{2}M_2'^2\right] \tag{8.18}$$

For $M_2' = 1$, Equation (8.18) reduces to

$$\left(\frac{M_1'}{1+\gamma M_1'^2}\right)^2\left[1+\frac{\gamma-1}{2}M_1'^2-\frac{\Delta H_R}{c_pT_1}\right]=\frac{1}{2(\gamma+1)} \tag{8.19}$$

Multiplying by $2(\gamma + 1)(1 + \gamma M_1'^2)^2/M_1'^2$:

$$2\gamma+2+(\gamma^2-1)M_1'^2+2(\gamma+1)-\frac{\Delta H_R}{c_pT_1}=\frac{1}{M_1'^2}+2\gamma+\gamma^2M_1'^2 \tag{8.20}$$

Simplifying and rearranging:

$$M_1'^2-2+\frac{1}{M_1'^2}=\frac{2(\gamma+1)(-\Delta H_R)}{c_pT_1} \tag{8.20a}$$

Taking the positive square root:

$$M_1'-\frac{1}{M_1'}=\left[\frac{2(\gamma+1)(-\Delta H_R)}{c_pT_1}\right]^{1/2} \tag{8.21}$$

The positive root of this quadratic in M_1' is

$$M_1'=\left[\left(\frac{\gamma+1}{2}\right)\left(\frac{-\Delta H_R}{c_pT_1}\right)\right]^{1/2}+\left[\left(\frac{\gamma+1}{2}\right)\left(\frac{-\Delta H_R}{c_pT_1}\right)+1\right]^{1/2} \tag{8.22}$$

As indicated in Fig. 8.9 the reaction may go only 50% to completion. Furthermore, the molar mass may change by a factor of 2 across the wave. Hence Equation (8.22) could only be expected to give qualitatively correct values. Even so, Morrison [16] has shown that the use of 2/3 of the value of ΔH_R for complete reaction, together with the initial value of γ and m gives values in reasonable agreement with experimental measurements for a variety of gaseous mixtures.

It follows from Equations (8.14) and (8.15) that

$$\frac{P_2}{P_1} = \frac{1 + \gamma M_1'^2}{1 + \gamma} \tag{8.23}$$

$$\frac{\rho_2}{\rho_1} = \frac{(1 + \gamma)M_1'^2}{1 + \gamma M_1'^2} \tag{8.24}$$

and

$$\frac{T_2}{T_1} = \left[\frac{1 + \gamma M_1'^2}{(1 + \gamma)M_1'}\right]^2 \tag{8.25}$$

Generalized, approximate expressions for shock and detonation waves

Adamson and Morrison [17] developed the following set of equations which reduce to the above approximate expressions for detonation waves when the coefficient F is equated to 1 and to Equations (7.15) – (7.18) for shock waves when F = 2:

$$\frac{P_2}{P_1} = 1 + \frac{\gamma F}{\gamma + 1}(M_1'^2 - 1) \tag{8.26}$$

$$\frac{u_2'}{u_1'} = \frac{\rho_1}{\rho_2} = 1 - \frac{F(M_1'^2 - 1)}{(\gamma + 1)M_1'^2} \tag{8.27}$$

$$\frac{T_2}{T_1} = \left[1 + \frac{\gamma F}{\gamma + 1}(M_1'^2 - 1)\right]\left[1 - \frac{F(M_1'^2 - 1)}{(\gamma + 1)M_1'^2}\right] \tag{8.28}$$

and

$$M_2'^2 = \frac{(\gamma+1-F)(M_1'^2-1)+\gamma+1}{F\gamma(M_1'^2-1)+\gamma+1} \qquad (8.29)$$

For detonation waves M_1' is estimated from (8.22).

Approximate relationships for reflected detonation waves

If no further reaction is assumed on reflection the previous expressions [(7.27) – (7.31)] for a shock wave are applicable but may be simplified by noting that $M_2 = 1$.

Complexities

Suggested sources of discussion of additional aspects are Sokolik [14], Emmons [18], Zel'dovich and Kompaneets [19] and Fay [20].

Problems

8.1. Set $\Delta H_{R1} = 0$ and drop the subscripts on the properties in Equation (8.4). Does this expression apply to shock waves?

8.2. Carry out the process in Problem 8.1 on Equation (8.13). Explain the result.

8.3. Drop the subscripts on the properties in Equation (8.13a). Does this result correspond to that of Equations (8.22) and (8.23)?

8.4. Compute P_2/P_1 and P_3/P_2 for a stoichiometric mixture of hydrogen and oxygen at 298 K and atmospheric pressure and compare the results with those in Table 8.1.

8.5. Using Fig. 8.11 – 8.13, test the assertion that the detonation and reflected pressures are proportional to the initial density.

8.6. A detonation wave is observed to travel at a velocity of 1220 m/s relative to the surroundings. Estimate the velocity of the (burned) gas behind the wave relative to the wall. An average heat capacity ratio of 1.3, an average molar mass of 0.024 kg/mol and an initial temperature of 289 K may be assumed. Estimate the effective value of $\Delta H_R/c_P$.

8.7. Estimate the maximum steady detonation wave pressure and impact pressure which might be developed if a detonation occurred in a stoichiometric mixture of CH_4 and O_2 at ambient conditions. Assume: a mean c_P of 41.9 J/mol·K, a negligible change in average molecular weight and $-\Delta H_R = 197.7$ kJ/mol of mixture for the equilibrium reaction

$$2O_{2\,(g)} + CH_{4\,(g)} \rightleftarrows CO_{2\,(g)} + 2H_2O_{(g)}$$

at existing conditions.

8.8. The velocity of a detonation wave in a gaseous mixture at 289 K and atmospheric pressure is determined by tests to be 1524 m/s. What pressure will be exerted on the equipment containing the mixture if a detonation occurs?
You may assume that $\gamma = 1.4$ and m = 0.030 kg/mol for both the initial mixture and the burned gas.

8.9. A gaseous process stream at 700 K and 0.7 MPa has a heat of combustion of approximately 4.2 MJ/kg and a mean specific heat capacity of approximately 1.25 kJ/kg·K. What maximum pressure will be encountered in the equipment if the mixture detonates?
It may be postulated that 2/3 of the heat of reaction will be released on detonation. Note any other significant assumptions which you make.

8.10. What is the maximum possible density ratio attainable across a shock wave; across a Chapman-Jouguet detonation wave? Explain the difference.

8.11. Does a shock wave or a detonation wave produce a higher pressure for the same Mach number (M_1')? Which produces a higher density? Explain briefly. Estimate the ratio of pressures and densities of the two waves, assuming the difference in physical properties can be neglected.

8.12. Will a detonation wave reflected by a wall generate a higher or lower impact pressure than a shock wave with the same initial velocity. Explain briefly.

8.13. A mixture of H_2 and O_2 is flowing from one tank at 0.70 MPa to another at 0.20 MPa through a short length of pipe. If a detonation occurs in the low pressure tank, will a detonation wave propagate through the pipe into the first vessel? Explain.

8.14. For the detonation of an ambient stoichiometric mixture of H_2 and O_2, assuming complete reaction to H_2O,
a) Prepare a plot of $(S_2 - S_1)/c_{pIJ}$ versus ρ_2/ρ_1.
b) Prepare a plot of T_2/T_1, u_2'/u_1', ρ_2/ρ_1, and M_2', M_1' versus P_2/P_1 for a complete range of P_2/P_1.
c) Show that S_2 goes through a maximum and a minimum at $M_2' = 0$.

8.15. Does the limiting case of $-\Delta H_R = 0$ in Equation (8.21) and (8.22) give the correct result for a shock wave? Explain.

8.16. Do the stagnation temperatures and pressures change across a shock wave and a detonation wave? Explain briefly.

8.17. Why is the reflected pressure rather than the pressure behind a shock or detonation wave used as a criterion for design?

8.18. Derive an expression for the "lost work" across a detonation wave.

References

1. I. Ginsburgh and W. L. Bulkley, "Hydrocarbon-Air Detonations–Industrial Aspects," *Chem. Eng. Progr.* **59**, No. 2, 82-86 (February, 1963).
2. J. O. Hirschfelder and C. F. Curtiss, "Introduction to the Theory of Detonations–I," University of Wisconsin Naval Research Laboratory Report under Contract NOrd 15834 with the Naval Bureau of Ordnance, Madison, Wisconsin (1957).
3. D. L. Chapman, "Rate of Explosion in Gases," *Phil. Mag.*, Ser. 5, **47**, 90-104 (1899).
4. E. Jouguet, "Sur la propagation des reactions chimiques dans le gaz," *J. de Math. Pures et Appliquees,* 6th Ser., **1**, 347-425 (1905), **2**, 1-86 (1906); *Mecanique des Explosifs*, Doin et fils, Paris (1917).
5. B. Lewis and G. von Elbe, *Combustion, Flames and Explosions of Gases*, Academic Press, New York (1951).
6. A. Weir, Jr. and R. B. Morrison, "Equilibrium Temperatures and Composition Behind a Detonation Wave," *Ind. Eng. Chem.* **46**, 1056-1060 (1954).
7. M. P. Moyle, R. B. Morrison and S. W. Churchill, "Detonation Characteristics of Hydrogen-Oxygen Mixtures," *AIChE Journal* **6**, 92-96 (1960).
8. R. L. Gealer and S. W. Churchill, "Detonation Characteristics of Hydrogen-Oxygen Mixtures at High Initial Pressures," *AIChE Journal* **6**, 501-505 (1960).
9. J. A. Luker, P. L. McGill and L. B. Adler, "Knallgas and Knallgas-Steam Mixtures at High Initial Temperature and Pressure," *Chem. Eng. Data* **4**, 136-143 (1959).
10. F. J. Zeleznik and S. Gordon, "An Analytical Investigation of Three General Methods of Calculating Chemical Equilibrium Compositions," NASA Tech. Note D-473, Washington, D.C. (September, 1960).
11. L. J. Zajak and A. K. Oppenheim, "Thermodynamic Computations for the Gas Dynamic Analysis of Explosion Phenomena," *Combustion and Flame* **13**, 537-550 (1969).
12. P. M. Ordin, "Hydrogen-Oxygen Explosions in Exhaust Ducting," NACA Tech. Note 3935, Washington, D.C. (April, 1957).
13. J. A. Luker and M. J. Leibson, "Dynamic Loading of Rupture Disks with Detonation Waves," *J. Chem. Eng. Data* **4**, 133-136 (1959).
14. Sokolik, A. S., *Self-Ignition, Flame and Detonation in Gases,* English translation, NASA and the Nat. Sci. Foundation, Washington, D.C. (1963).
15. R. N. Randall, J. Bland, W. M. Dudley and R. B. Jacobs, "The Effects of Gaseous Detonations Upon Vessels and Piping," *Chem. Eng. Progr.* **53**, No. 12, 574-580 (December, 1957).
16. R. B. Morrison, "A Shock Tube Investigation of Detonative Combustion," Report AF 33 (038) 12657, UMM-97, Willow Run Research Center, University of Michigan, Ann Arbor (1952).

17. T. C. Adamson, Jr. and R. B. Morrison, "On the Classification of Normal Detonation Waves," *Jet Propulsion* 25, 400, 403 (1955).
18. H. W. Emmons., *Fundamentals of Gas Dynamics,* Princeton University Press, Princeton, New Jersey (1958).
19. Ya. B. Zel'dovich and A. S. Kompaneets, *Theory of Detonation,* English trans., Academic Press, New York (1960).
20. J. A. Fay, "Two-Dimensional Gaseous Detonations: Velocity Deficit," *Phys. Fluids* 2, 283-289 (1959).

Chapter 9

Surface waves

Surface waves in liquids are generated by gravity and surface tension. The surface waves generated by gravity are somewhat analogous to shock waves in gas. Gravitational waves are of particular industrial importance because they provide an upper limit for the rate of flow in open channels, just as acoustic waves provide an upper limit in closed channels. Such waves will be considered herein. The waves generated by surface tension are called capillary waves. They will be considered only briefly, although they are of industrial importance in thin films.

Gravitational waves

Consider a liquid flowing at velocity u_1 and depth d_1 in an open rectangular channel of width s. A disturbance downstream is presumed to cause a surface wave to propagate upstream at a steady velocity u_w, resulting in an increased depth d_2 and a decreased velocity u_2 downstream, as shown in Fig. 9.1.

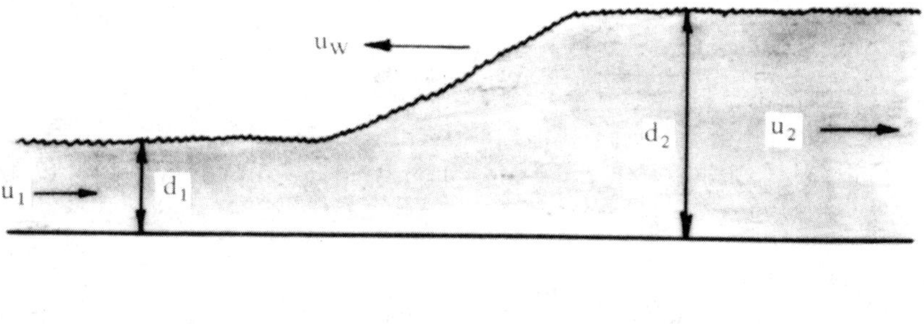

Fig. 9.1. Moving gravitational wave

All of the velocities will be considered uniform as a first-order approximation. Changing the frame of reference to that of the wave yields the stationary process shown in Fig. 9.2.

A material balance for the stationary representation takes the form

$$u_1' \rho s d_1 = u_2' \rho s d_2 \tag{9.1}$$

If pressure is considered to be the only force, i.e., if viscosity, surface tension, etc. are neglected, a momentum and force balance can be written as follows:

$$sP_0(d_2 - d_1) + u_1'^2 \rho s d_1 + s\int_0^{d_1} P dy = u_2'^2 \rho s d_2 + s\int_0^{d_2} P dy \tag{9.2}$$

where P_0 is the atmospheric pressure on the surface and y is the distance from the surface of the liquid. The pressure in the liquid can be approximated by the static pressure.

Hence

$$P = g\rho y + P_0 \tag{9.3}$$

and the difference of the integrals in Equation (9.2) is

$$\int_0^{d_2} P\,dy - \int_0^{d_1} P\,dy = g\rho\left(\frac{d_2^2 - d_1^2}{2}\right) + P_0(d_2 - d_1) \tag{9.4}$$

Combining Equations (9.2) and (9.4):

$$sg\rho\left(\frac{d_2^2 - d_1^2}{2}\right) = (u_1'^2 d_1 - u_2'^2 d_2)s\rho \tag{9.5}$$

Cancelling out $s\rho$ from both sides and eliminating u_2 through Equation (9.1) gives

$$u_1'^2 d_1 - \frac{u_1'^2 d_1^2}{d_2} = \frac{g}{2}(d_2^2 - d_1^2) \tag{9.6}$$

which can be rearranged as

$$u_1'^2 = \frac{gd_2}{2}\left(1 + \frac{d_2}{d_1}\right) \tag{9.7}$$

For a small disturbance such that

$$d_2 \cong d_1 = d \tag{9.8}$$

$$u_1'^2 \cong gd \tag{9.9}$$

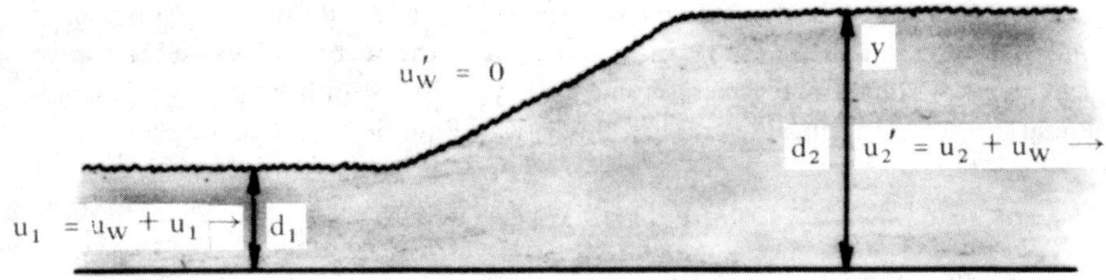

Fig. 9.2. Stationary gravitational wave

From Equation (9.7), Fig. 9.2 and Equation (9.9)

$$u_W = [gd_2(d_1 + d_2)/2d_1]^{1/2} - u_1 \cong (gd)^{1/2} - u_1 \tag{9.10}$$

Thus if u_1 is much less than $(gd)^{1/2}$, $u_W > 0$ and the wave will propagate upstream. If u_1 is much greater than $(gd)^{1/2}$, $u_W < 0$ and the wave will be swept downstream.

The irreversibilities due to the formation of a wave can be determined by rewriting Equation (6.5) (see Problem 9.20) in terms of the frame of reference of a stationary wave as

$$lw_{12} = g(d_1 - d_2) + \frac{u_1'^2 - u_2'^2}{2} \tag{9.11}$$

Eliminating u_1' and u_2' through Equations (9.1) and (9.7) gives on simplification

$$lw_{12} = g(d_2 - d_1)^3/4d_2 d_1 \tag{9.12}$$

The "lost work" can also be expressed in terms of u_1 and d_1 instead of d_1 and d_2 (see Problem 9.21).

The dimensionless group u^2/gd is called the Froude number* and is usually symbolized by Fr. This group can be interpreted as the ratio of the inertial force per unit area ($u^2\rho$) to the gravitational force per unit area ($g\rho d$) and is obviously the criterion for the propagation of surface waves due to gravity.

**After William Froude who discovered its significance in connection with the design of ships [1].*

Hydraulic jump

A stationary wave in an open flowing stream of liquid is called a *hydraulic jump*. Fig. 9.2 can be considered to represent a hydraulic jump. (The use or omission of primes is arbitrary in this case since the surroundings and the wave provide the same frame of reference. They will be omitted for simplicity). Equation (9.7) is applicable and it follows from Equation (9.1) that

$$u_2^2 = gd_1(d_1 + d_2)/2d_2 \tag{9.13}$$

From (9.7) and (9.13) it can be concluded that if

$$d_2 > d_1,\quad u_1 > (gd_2)^{1/2} > (gd_1)^{1/2} \text{ and } u_2 < (gd_1)^{1/2} < (gd_2)^{1/2}$$

$$d_2 = d_1,\quad u_1 = u_2 = (gd_1)^{1/2} = (gd_2)^{1/2}$$

$$d_2 < d_1,\quad u_1 < (gd_2)^{1/2} < (gd_1)^{1/2} \text{ and } u_2 > (gd_1)^{1/2} > (gd_2)^{1/2}$$

Thus the minimum upstream velocity for hydraulic jump is $(gd_1)^{1/2}$. For higher velocities the fluid will jump spontaneously yielding a stable velocity u_2 less than $(gd_2)^{1/2}$. The downstream conditions u_2 and d_2 can be calculated from the upstream conditions u_1 and d_1 by solving Equation (9.7) for d_2. The positive root is

$$d_2 = d_1([1 + (8u_1^2/gd_1)]^{1/2} - 1)/2 \tag{9.14}$$

and from Equation (9.1)

$$u_2 = \frac{u_1 d_1}{d_2} = \frac{2u_1}{\left(1 + \dfrac{8u_1^2}{gd_1}\right)^{1/2} - 1} \tag{9.15}$$

Velocities greater than $(gd_1)^{1/2}$ may occur, e.g., due to an increase in slope or a decrease in the width of a channel, or to flow under a gate, as illustrated in Fig. 9.3.

The subsequent shift to $u_2 < u_1$ and $d_2 > d_1$ sets the minimum required height of the retaining walls downstream. The flow for $u_2 < (gd)^{1/2}$ is called *streaming*, and that for $u_1 > (gd)^{1/2}$ *shooting*.

An experimental test of the applicability of Equation (9.7) and hence (9.13) – (9.15) to hydraulic jump is provided in Fig. 9.4 from Woodward and Posey [2]. Here the diameter ratio d_2/d_1 across the jump is plotted versus the upstream Froude number $u_1^2/2gd_1$. Good agreement is observed with Equation (9.7). An expression for reversible flow can be obtained by setting lw_{12} equal to

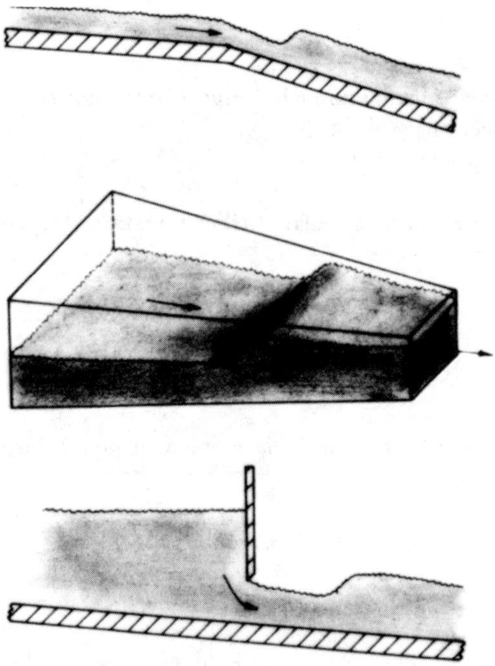

Fig. 9.3. Three conditions producing a hydraulic jump

zero in Equation (9.11) and eliminating u_2' through (9.1) to obtain

$$\left(\frac{u_1^2}{2gd_1}\right)_{rev} = \frac{\dfrac{d_2}{d_1}}{1+\dfrac{d_1}{d_2}} \tag{9.16}$$

Equation (9.16) is plotted in Fig. 9.4 as an upper bound and hence a test of the consistency of the data.

Maximum rate of flow in an open channel

A Froude number of unity defines the maximum stable rate of flow through an open channel since a higher velocity would lead to hydraulic jump and overflow. The slope of the channel required to yield any rate of flow, including this limiting value, can be calculated approximately from Equation (5.1) after modification as follows:

$$\tau p_w \, dx - g\rho \sin\{\theta\} A \, dx = 0 \tag{9.17}$$

where here

θ = the angle of inclination with respect to the horizontal, rad
A = cross-sectional area of fluid, m^2
p_w = wetted perimeter of channel, m

For turbulent flow τ can be calculated from Equations (5.2) and (5.3) or (5.4) using the equivalent diameter

$$D_e \equiv 4A/p_w \qquad (9.18)$$

Surface waves in a stagnant liquid

The velocity of a surface wave in a stagnant liquid due to gravity is (see, e.g., Kaufmann [2])

$$u_w = [(g\lambda/2\pi) \tanh\left\{2\pi d/d\right\}]^{1/2}$$

where λ = the wavelength, m

For long waves in shallow water ($\lambda \gg d$), $d/\lambda \to 0$, $\tanh\left\{2\pi d/\lambda\right\} \to 2\pi d/\lambda$ and

$$u_w \to (gd)^{1/2} \qquad (9.20)$$

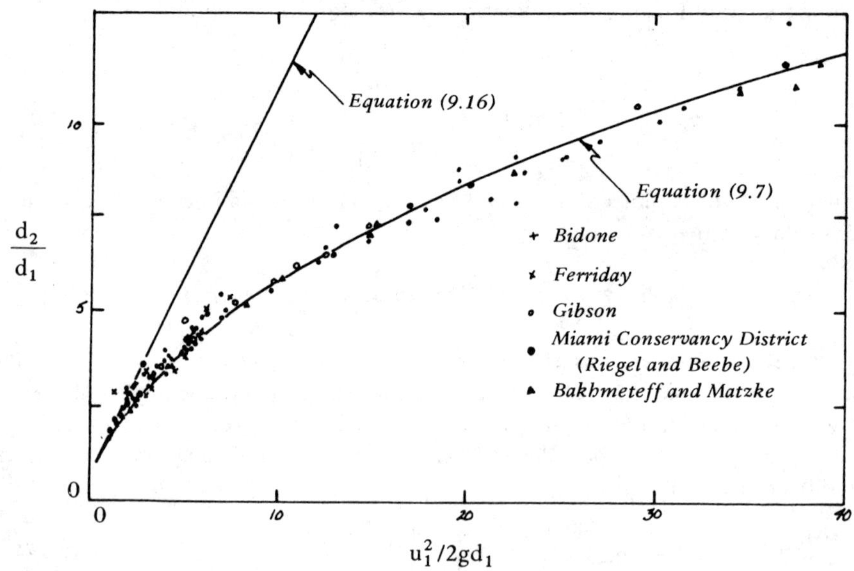

Fig. 9.4. Experimental test of theoretical expression for hydraulic jump [2]

in accord with Equation (9.9). The velocity in this limit is independent of wavelength and depends only on depth.

For short waves in deep water ($\lambda \ll d$), $\lambda/d \to 0$, $\tanh\{2\pi d/\lambda\} \to 1$ and

$$u_W \to (g\lambda/2\pi)^{1/2} \tag{9.21}$$

Thus in deep water long waves propagate more rapidly than short waves. This dependence on wavelength is analogous to dispersion in optics and leads to the complex wave patterns observed on large bodies of water. The period of a wave is by definition

$$T = \lambda/u_W \tag{9.22}$$

and hence in deep water

$$T \to \lambda(2\pi/g\lambda)^{1/2} = (2\pi\lambda/g)^{1/2} \tag{9.23}$$

The effect of surface tension on the wave velocity in deep water ($\lambda/h \ll 2\pi$) can be represented by

$$u_W = \left[\frac{g\lambda}{2\pi} + \frac{2\pi\sigma}{\lambda\rho}\right]^{1/2} \tag{9.24}$$

where σ = surface tension, N/m or kg/s^2.

For $\lambda/\sigma \to 0$, pure capillary waves are attained with a velocity

$$u_W = (2\pi\sigma/\lambda\rho)^{1/2} \tag{9.25}$$

For $\sigma/\lambda \to 0$ Equation (9.24) reduces to (9.21) for gravity waves.

Complexities

More detailed treatments of surface waves and open channel flow are provided by books on hydraulics, such as those by Kaufmann [3], Chou [4], and Stokes [5].

Problems

9.1. Estimate the velocity of a long wave in a 1-m-deep bay. Estimate the velocity of ocean waves spaced 12 m between crests.

9.2. An open channel overflows its sides. It is proposed to correct this difficulty by increasing the depth or the width. Which is more effective?

9.3. What is the maximum rate of flow in kg/s which can be achieved in a rectangular ditch 1.22 m wide and 0.9 m deep? How is this rate affected by the slope and roughness?

9.4. Calculate the slope of a 0.6-m-deep, 1.8-m-wide channel through which water will flow at critical velocity.

9.5. What is the capacity in m^3/s of a semi-circular concrete ditch 2 m in diameter. What slope is required?

9.6. You are considering wading through a small stream but can't see the bottom because of the turbidity of the water. Fortunately, you observe a hydraulic jump of 0.2 m in the stream, and from floating chips of wood you estimate the downstream velocity to be 2.1 m/s. What is the downstream depth?

9.7. A wave propagating through stagnant water encounters a wall and is reflected. The stagnant water is only 0.30 m deep at the wall and height of the wave itself is 0.9 m (1.2 m above the bottom).

a) What is the velocity of the original wave relative to the wall?

b) What is the velocity of the water behind the original wave relative to the wall?

d) What is the height of the reflected wave above the bottom?

9.8. Two students observe a standing wave downstream from the sluice gate in Fig. 9.3. One asserts he can calculate the flow rate from the depth of water before and after the wave. The other says he can predict the flow rate from only the height of water behind the dam and height of the gate. Which student is correct? Justify your answer by writing out the required relationships.

9.9. What minimum height is required for the walls of the spillway downstream from the sluice gate in Problem 2.5?

9.10. An indoor surfing tank is to generate a wave with a velocity of 4.5 m/s over a stagnant pool 1.0 m deep. How high should the wave be?

9.11. An indoor surfing tank generates a 1-m-high wave above a stagnant pool 1.33 m deep. Calculate the velocity of the wave in m/s.

9.12. A wave 0.15 m high is observed to travel up a river at a velocity of 0.9 m/s relative to the river bank. What is the velocity of the river?

9.13. A rectangular ditch 1.2 m wide and 0.9 m deep is used where the slope is 0.75 m/km. What dimensions should be used for a continuation of this ditch in a section where the slope is 2.24 m/km?

9.15. Water is flowing through a rectangular ditch 0.9 m deep and 1.2 m wide. How much would the ditch need to be deepened to double the volumetric rate of flow? State and justify any assumptions.

9.16. A rectangular channel 2 m wide gradually decreases in width. If the rate of flow of water in the 2-m section is 3 m^3/s and the depth 1 m, at what width and depth will the Froude number reach unity?

9.17. A water wave is observed to reflect from a wall at a height of 0.3 m above the approaching water which has a depth of 1.2 m. What is the velocity of the wave and approaching water relative to the wall.

9.18. The flow pattern on an experimental sieve tray operating at a high liquid rate and with an insoluble gas is shown in the diagram. The liquid flows from left to right across the tray. A–B is a region of clear liquid, in which the velocity is u_L and the density is ρ_L. The perforations begin at B, and B–C is a froth zone of height d_F. Stating any necessary assumptions, show that the bulk density ρ_F of the froth is given by the solution of

$$\left(\frac{d_F}{d_L}\right)^3 \frac{\rho_F}{\rho_L} - \frac{d_F}{d_L}\left[1 + \frac{2u_L^2}{gd_L}\right] + \frac{2u_L^2\rho_L}{gd_L\rho_F} = 0$$

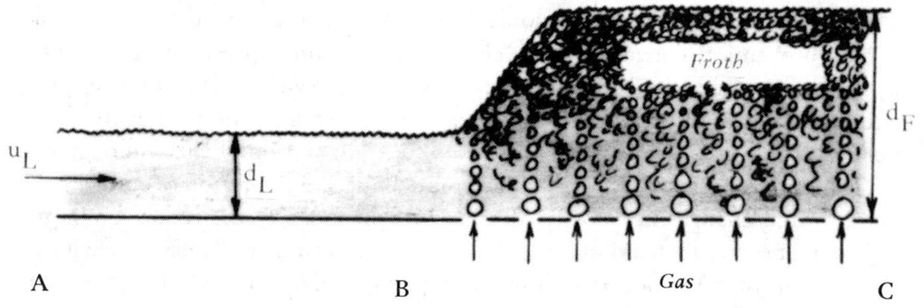

Fig. 9.5. Schematic drawing of an experimental sieve tray

9.19. A standing wave 180 mm high is observed in a river flowing at 5 m/s. Estimate the depth of the river assuming the wave was caused by a change in the slope of the bottom rather than by an obstruction.

References

1. W. Froude, "Experiments with Models Capable of Application to Full-Sized Ships," *Trans. Inst. Naval Arch.* **11**, 88-89 (1870).
2. S. M. Woodward and C. J. Posey, *Hydraulics of Steady Flow in Open Channels*, Wiley, New York (1941), p. 29.
3. W. Kaufmann, *Fluid Mechanics*, English transl. by E. G. Chilton, McGraw-Hill, New York (1963).
4. K. T. Chou, *Open-Channel Hydraulics*, McGraw-Hill, New York (1959).
5. J. J. Stokes, *Water Waves*, Interscience Publishers, New York (1957).

Chapter 10

Cavitation, incipient vaporization and aerodynamic heating

High velocities produce a number of special effects besides those noted in previous chapters. Several of these are considered herein. *Cavitation* and *incipient vaporization* often constitute a limiting condition of flow. *Aerodynamic heating* may provide a restriction on the allowable rate of flow.

Cavitation

If the pressure of a flowing liquid falls below the vapor pressure, bubbles may form; if the pressure increases again these bubbles may recondense or redissolve. The *rapid* formation and collapse of such bubbles is known as cavitation. This process may produce vibrations, shock waves, operational disorders, erosion and even chemical corrosion. (The chemical corrosion is presumed to be due to hydrogen ions generated by the shock waves). Dissolved air may reduce the deleterious effects of cavitation since the formation and re-solution of the air bubbles is slower than the vaporization and recondensation of the liquid itself.

The rapid pressure changes which produce severe cavitation occur primarily due to rapid changes in velocity although the necessary gross reduction in pressure may be due to friction and changes in elevation as well. Severe cavitation therefore most often occurs in flow through nozzles, in flow over hydrofoils and propellers, and in flow through pumps and turbines where rapid changes in velocity are inherent. Because of this dependence on velocity, cavitation may constitute an effective upper limit for the flow of liquids through piping systems and an important consideration in the design of propellers, impellers, turbine blades and hydrofoils.

a) Nozzles. For the reversible flow of a liquid through a nozzle, Equation (1.2) can be written in terms of a general P and ρ as

$$P_1 - P = \rho(u^2 - u_1^2)/2 + g\rho(z - z_1) \qquad (10.1)$$

where here the subscript 1 indicates either inlet or exit conditions. Cavitation may occur if the pressure at any point falls below the vapor pressure of the liquid stream, hence if

$$P = P_1 - \rho(u^2 - u_1^2)/2 - g\rho(z - z_1) < \mathbf{P} \qquad (10.2)$$

where

$\mathbf{P}$ = saturated vapor pressure of the liquid, Pa

For the horizontal flow of a liquid, the maximum velocity and minimum pressure occur at the throat of a converging-diverging nozzle as well as at the throat of a converging nozzle. Hence the condition for incipient cavitation becomes

$$P_t = P_1 - \rho(u_t^2 - u_1^2)/2 = \mathbf{P} \tag{10.3}$$

The gravitational term is ordinarily negligible relative to the velocity even in a vertical nozzle, hence Equation (10.3) is a good approximation for all conditions.

b) Pumps. The pressure at the inlet of a pump transferring liquid from a vessel as illustrated in Fig. 10.1 is

$$P_i = P_1 + g\rho d - \Delta P_f \tag{10.4}$$

where here

P_i = pressure at inlet of pump, Pa
P_1 = pressure on the surface of the liquid, Pa
d = elevation of the liquid level above the pump inlet, m
$-\Delta P_f$ = pressure drop due to friction and contraction, Pa

Equation (10.4) can be applied at any point in the piping system if P_1, d and ΔP_f are referred to that point.

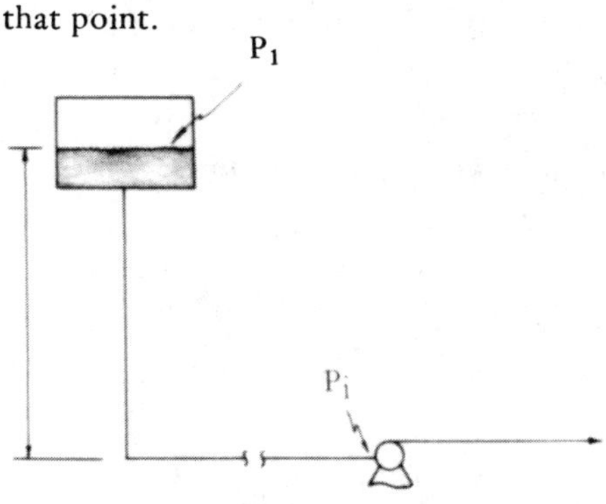

Fig. 10.1. Pumping from a vessel

The maximum decrease in pressure inside a centrifugal pump due to high local velocities is expressed in *head units** and is called the *net positive suction head (NPSH)* of the pump. This particular designation is used in the pump literature because the equivalent pressure is the positive pressure increment above the vapor pressure, which must exist at the inlet if cavitation is to be avoided. The net positive suction head varies somewhat with the throughput of the pump but manufacturers usually specify a single conservative value.

Thus to avoid cavitation it is necessary that

$$P_1 + g\rho d - \Delta P_f - g\rho(\text{NPSH}) > P \qquad (10.5)$$

Since ΔP_f is a function of the velocity in the piping system, cavitation provides an upper limit for the rate of pumping.

If P_1 is equal to the vapor pressure over the liquid, as is the case for the condensate from the overhead of a distillation column, as illustrated in Fig. 10.2, the condition for incipient cavitation is

$$g\rho d - \Delta P_f = g\rho(\text{NPSH}) \qquad (10.6)$$

Cavitation in a centrifugal pump produces rattling, vibration and a sharp drop in the efficiency and rate of pumping. If cavitation is permitted to continue the pump may be damaged.

c) Flow over immersed surfaces. The acceleration of liquids in flow over propellers and hydrofoils may also produce cavitation and therefore provides a limiting condition for their operation. Knapp and Hollander [1] took high-speed photographs (20,000 frames per second) of water flowing over hydrofoils and found that vapor bubbles grew to 8 mm in 2 ms and collapsed in 1 ms. On her maiden voyage the ill-fated ocean-liner *Lusitania* lost 50 mm of metal from the surface of the propellers due to cavitation.

**Equation (1.1) can be rearranged in the following form*

$$\frac{dP}{\rho g} + dz + \frac{udu}{g} = 0 \qquad (1.1a)$$

or for constant ρ and g as

$$d\left(\frac{P}{\rho g} + z + \frac{u^2}{2g}\right) = 0 \qquad (1.1b)$$

which is known as the head form of Bernoulli's equation. The pressure and velocity terms are thus converted to equivalent elevations or heads. The term $P/\rho g$ is called the pressure head, the term z the gravitation head and the term $u^2/2g$ the velocity head. These terms have the units of $Pa{\cdot}m^2{\cdot}s^2/kg$ or m. The pressure loss due to friction and the pressure increase generated by a pump can similarly be expressed as heads simply by dividing by ρg.

Units of head are sometimes advantageous when working with liquids in that the changes can be interpreted directly in terms of elevation. In addition the work done by a centrifugal pump is approximately the same for all fluids when expressed in units of head.

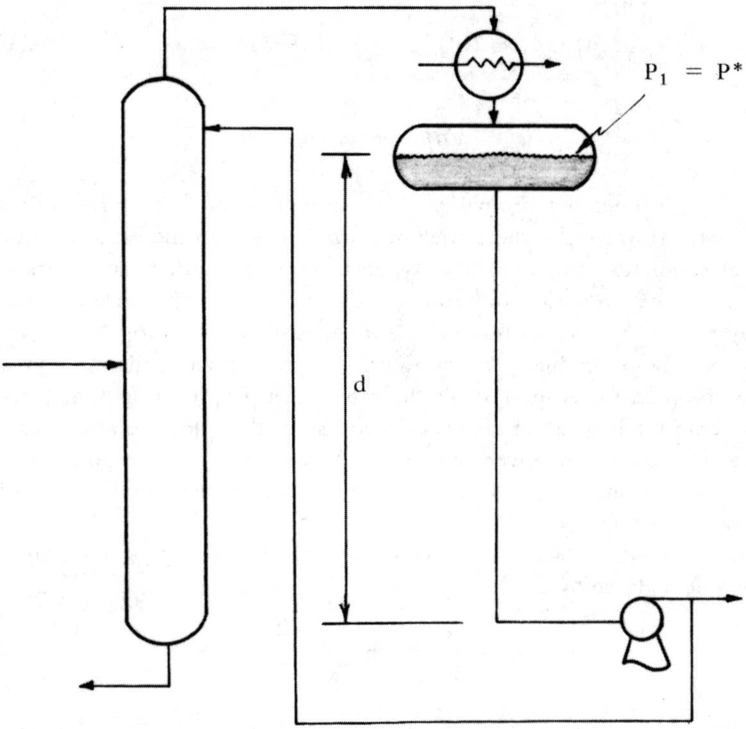

Fig. 10.2. Pumping saturated liquid from an overhead accumulator

Equation (10.1) can be reduced to the following form for a hydrofoil or propeller by assuming that $u_1 = 0$ and that P_1 is the atmospheric pressure:

$$P_{atm} - P_{min} = \rho u_{max}^2 /2 - g\rho d \qquad (10.7)$$

where here

P_{min} = minimum pressure, Pa or kg/m·s^2
u_{max} = maximum velocity on surface of hydrofoil or propeller, m/s
d = depth below surface, m

Hence, the condition for incipient cavitation is

$$P_{atm} - \rho u_{max}^2 /2 + g\rho d = P \qquad (10.8)$$

or

$$u_{max}^2 = 2(P_{atm} - P)/\rho + 2gd \tag{10.8a}$$

Vapor-binding

Cavitation does not ordinarily occur in pipes because the pressure usually falls slowly. However, if the outlet pressure is less than the vapor pressure, or if the elevation at some point is sufficiently above the outlet, vaporization may take place. The formation of vapor greatly increases the pressure gradient and thereby reduces the rate of flow for a given overall pressure drop. This condition is known as vapor-binding. Reduction of the pressure below the vapor pressure, due to the pressure drop through the valves of a reciprocating pump, usually causes vapor-binding rather than cavitation since the velocities and rates of re-compression are much lower than in centrifugal pumps. The point at which vaporization begins can be calculated by extension and modification of the above results for nozzles.

For constant density the velocity does not vary in a pipe. Equation (5.1) can then be replaced by

$$\tau\pi D\,dx + \frac{\pi D^2}{4}\,dP + g\rho\,\frac{\pi D^2}{4}\,dz = 0 \tag{10.9}$$

Replacing τ with f from Equation (5.2) and rearranging:

$$f\,dx = \frac{-D\,dP}{4\rho u^2} - \frac{gD\,dz}{4u^2} \tag{10.10}$$

Integrating for constant f from x = 0 to x, $P = P_1$ to P, and z = 0 to d gives

$$fx = \frac{D(P_1 - P)}{4\rho u^2} - \frac{gDd}{4u^2} \tag{10.11}$$

where here

x = length of pipe from the outlet, m
d = elevation above the outlet, m
P_1 = pressure at the outlet, Pa

The criterion for incipient vaporization is then

$$P_1 - g\rho d - \frac{4f\rho u^2 x}{D} = \mathbf{P} \tag{10.12}$$

and the point of occurence is

$$x = \frac{D(P_1 - \mathbf{P})}{4\rho f u^2} - \frac{gdD}{4fu^2} \tag{10.12a}$$

The friction factor can be calculated conveniently from Equation (5.4) in this case, and then x. In general, d will be a function of x.

Equation (10.12) can be modified to determine the maximum rate of pumping in the situation of Fig. 10.3 by including the NSPH of the pump, replacing x with the total equivalent length of the piping system **L**, letting d be the elevation of the pump above the pool and letting P_1 be the atmospheric pressure. Then

$$P_1 - g\rho d - \frac{4f\rho u_{max}^2 \mathbf{L}}{D} - g\rho(\text{NPSH}) = \mathbf{P} \tag{10.13}$$

or

$$u_{max}^2 = \frac{D(P_1 - \mathbf{P})}{4\rho f\mathbf{L}} - \frac{gdD}{4f\mathbf{L}} - \frac{gD(\text{NPSH})}{4f\mathbf{L}} \tag{10.13a}$$

Equation (10.13) can be recognized as the equivalent of Equation (10.5) except for the reversed sign on the gravitational term, owing to the definition of d as the elevation of the pump above the liquid level in Equation (10.13) and as the converse in Equation (10.5).

Effective vapor pressure

Some judgement is required in evaluating the vapor pressure **P**, as used in different applications in this chapter. In the case of a pure liquid, **P** is the equilibrium vapor pressure. For small bubbles the vapor pressure depends on the diameter and surface tension. If a gas is dissolved in the liquid, such as air in water, the dissolved gas contributes to the vapor pressure. As noted above, evolution and re-solution of dissolved gas occurs more slowly than vaporization of the liquid and does not contribute as strongly to the deleterious effects of cavitation. However, to be conservative, the partial pressure of the dissolved gas may be added to the vapor pressure of the pure liquid in calculating the conditions

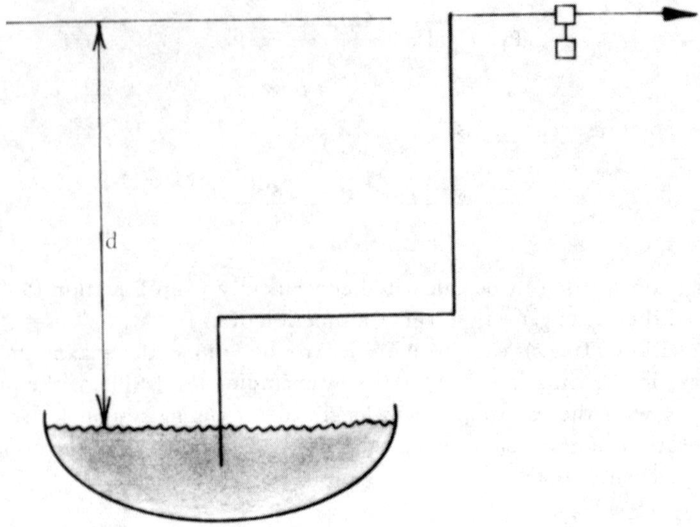

Fig. 10.3. Pumping liquid from an underground pool

for incipient cavitation. In the case of vapor-binding it is appropriate to calculate both the onset of gas evolution and of vaporization since the accumulation of evolved gas at some point in the system may have undesirable effects. In the case of an exposed pool, as in Fig. 10.3, the liquid will be saturated with the ambient gas at the ambient pressure, and dissolution will occur upon any reduction in pressure.

Aerodynamic heating

If a gas stream is slowed down by flowing over or impacting on a surface, or by expanding into a large reservoir, the temperature increases due to the conversion of kinetic energy to thermal energy. This phenomenon is usually called aerodynamic heating. For high initial velocities the temperature rise may be very great. Indeed, at one time supersonic flight was considered impossible for this reason. (The motion of an airplane through stagnant air produces exactly the same effect as air motion over a fixed airplane). The temperature rise of a thermometer due to aerodynamic heating may be significant even at moderate velocities.

For free, adiabatic, horizontal flow Equation (1.16) reduces to

$$H - H_1 + \frac{u^2 - u_1^2}{2} = 0 \tag{10.14}$$

where here the subscript 1 refers to any reference condition. Equation (10.14) is of course identical to Equation (7.3) but here slow changes in area will be allowed, as long as the flow can still be treated as one-dimensional. For an ideal gas with constant heat capacity and no reaction, Equation (7.4) is applicable, and combination with Equation (10.4) leads to Equation (1.44) which was previously derived for the unnecessarily restricted case of reversible flow. Equation (1.45) for the stagnation temperature can be rearranged as

$$\frac{T_s}{T_1} = 1 + \frac{\gamma - 1}{2} M_1^2 \tag{10.15}$$

Equation (10.15) is applicable to flows with or without friction, to nozzles with varying cross section and to pipes with constant cross section.

In nozzles the flow is essentially adiabatic and reversible, and the change in velocity results from the change in area. If ideal gas behavior and a constant heat capacity are assumed, Equations (1.38) and (10.15) can be combined to calculate the corresponding changes in pressure and density. Thus

$$\frac{P_s}{P_1} = \left[1 + \frac{\gamma - 1}{2} M_1^2\right]^{\frac{\gamma}{\gamma - 1}} \tag{10.16}$$

and

$$\frac{\rho_s}{\rho_1} = \left[1 + \frac{\gamma - 1}{2} M_1^2\right]^{\frac{1}{\gamma - 1}} \tag{10.17}$$

From Equations (10.16) and (10.17) it is apparent that the stagnation pressure and density are invariant in adiabatic reversible flow of an ideal gas with a constant heat capacity. An infinite cross-sectional area would be required to attain T_S, P_S, ρ_S and zero velocity in a nozzle. Hence these values should be considered as the limiting condition which is approached as the area increases indefinitely, either upstream or downstream.

In adiabatic flow through a pipe with constant cross section, any change in velocity must be due to friction. For an ideal gas and a constant cross section the derivation beginning with Equation (5.13) can be repeated with Equation (5.14) simplified to

$$\frac{u_1^2}{2} = \frac{G^2}{2\rho_1^2} = \frac{\gamma}{\gamma - 1}\left(\frac{P_1}{\rho_1} - \frac{P_s}{\rho_s}\right) \tag{10.18}$$

by setting u = 0. Then Equation (5.17) is replaced by

$$\frac{8fL}{D} = \ln\left\{\frac{\rho_s}{\rho_1}\right\}^2 + \frac{1}{\gamma}\left[\frac{1}{M_1^2} + \frac{\gamma-1}{2}\right]\left[1 - \left(\frac{\rho_s}{\rho_1}\right)^2\right] \tag{10.19}$$

The stagnation pressure can be calculated from Equation (1.12) using T_S from Equation (10.15) and ρ_S from Equation (10.19). P_S and ρ_S should be considered as hypothetical limiting values for pipe flow. The stagnation conditions are necessarily upstream from State 1.

An adiabatic surface exposed to a flowing stream does not attain exactly the stagnation temperature owing to viscous heating and convective and radiative cooling. If radiation is neglected the *adiabatic surface temperature*, T_{aw}, can be correlated in terms of a *recovery factor* r, defined as

$$r \equiv \frac{T_{aw} - T_1}{T_s - T_1} = \frac{T_{aw} - T_1}{u_1^2/2c_p} \tag{10.20}$$

Eckert and Drake [2] recommend the following expressions for the recovery factor in boundary layers:

laminar	$r \cong Pr^{1/2}$	(10.21)
turbulent	$r \cong Pr^{1/3}$	(10.22)

where

$Pr = \nu/a$ = Prandtl number

ν = kinematic viscosity, m^2/s

a = thermal diffusivity, m^2/s

Combination of Equations (10.15) and (10.20) leads to

$$T_{aw}/T_1 = 1 + r(\gamma - 1)M_1^2/2 \tag{10.23}$$

For $Pr < 1$, hence for most gases, the recovery factor is less than unity and the adiabatic surface temperature is less than than the stagnation temperature,

i.e., convective cooling predominates over viscous heating. For $Pr > 1$, the recovery factor is greater than unity and the adiabatic surface temperature is greater than the stagnation temperature, i.e., viscous heating predominates over convective cooling.

For non-adiabatic surfaces the rate of convective heat transfer can be represented by

$$j = h(T_W - T_{aw}) \tag{10.24}$$

where

j = convective heat flux density from surface, W/m^2 or kg/s^3 or $J/m^2 \cdot s$
T_W = surface temperature, K
h = heat transfer coefficient, $W/m^2 \cdot K$ or $kg/s^3 . K$ or $J/m^2 \cdot s \cdot K$

The coefficient h can be calculated from the same correlations as for low-velocity flow. Equation (10.23) indicates that the fluid stream behaves thermally as if it were at T_{aw} instead of T_1. Equations (10.19) – (10.23) are applicable to flow inside tubes if T_1 is replaced by T_b, the mixed-mean temperature.

The flowing temperature fields for air at high speeds over heated, cooled and adiabatic surfaces are shown schematically in Fig. 10.4. The subscript ∞ indicates the free-stream velocity and temperature far from the surface. The local stagnation and flowing temperature fields for flow at high speeds over an adiabatic surface are similarly compared in Fig. 10.5. If the plate is adiabatic it is necessary that the mean stagnation temperature of the air equals the free-stream stagnation temperature $T_{\infty s} = T_\infty + u_\infty^2/2c_p$. Since the adiabatic wall temperature for a gas is necessarily less than $T_{\infty s}$, the local stagnation temperature must exceed T_{aw} and even $T_{\infty s}$ over a short range of distances as indicated.

If radiation from surroundings of infinite extent at uniform temperature T_0 is also taken into account

$$j = h(T_W - T_{aw}) + \epsilon\sigma[T_w^4 - T_0^4] \tag{10.25}$$

where here

j = net heat flux density, W/m^2
ϵ = emissivity of surface
σ = Stefan coefficient = 5.6698×10^{-8} $W/m^2 \cdot K^4$ or $kg/s^3 \cdot K^4$ or $J/s \cdot m^2 \cdot K^4$
T_0 = temperature of surroundings, K

If the surface is adiabatic

$$(T_W - T_{aw})h = \epsilon\sigma[T_0^4 - T_w^4] \qquad (10.26)$$

Substituting for T_{aw} from Equation (10.20) gives

$$\left(T_W - T_1 - \frac{ru_1^2}{2c_P}\right)h = \epsilon\sigma[T_0^4 - T_W^4] \qquad (10.27)$$

and on rearrangement

$$T_W = T_1 + ru_1^2/2c_P + \epsilon\sigma(T_0^4 - T_W^4)/h \qquad (10.27a)$$

Equation (10.27a) gives the surface temperature T_W, as a function of the flowing gas temperature T_1. It can be applied conversely to give the temperature of a gas stream in terms of the temperature of a thermometer or thermocouple in the stream of gas. A more detailed treatment of aerodynamic heating is provided by Truitt [3].

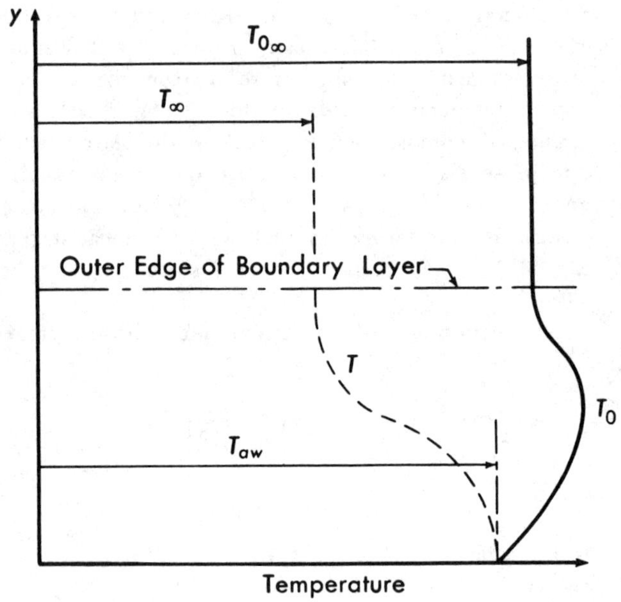

Fig. 10.4. Flowing and stagnation temperature fields for flow of air over heated, cooled and adiabatic surfaces [3]

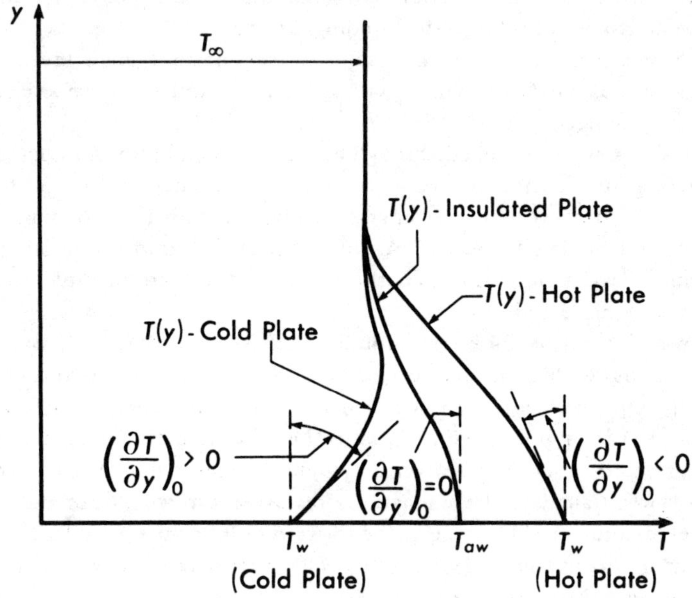

Fig. 10.5. Flowing and stagnation temperature fields for flow of air over an adiabatic surface [3]

Problems

10.1. At what rate of flow, if any, will cavitation occur in 340 K, air-free water discharging to the atmosphere through a converging-diverging nozzle 1.524 m-long, 100 mm in diameter at the exit and 50 mm in diameter at the throat?

10.2. What is the maximum rate at which butane can be pumped from an accumulator at 311 K by a centrifugal pump with a NPSH of 2.44 m? The liquid level in the accumulator is 0.61 m. The 102 mm-suction line descends 3.66 m below the accumulator, runs 9.1 m horizontally, rises 0.9 m and then runs 1.2 m horizontally to the pump suction. There is one gate valve at the accumulator outlet and another at the pump suction; an equivalent length of 2.13 m may be assumed for each.

10.3. Estimate the maximum temperature attained by the surface of an airplane flying at 225 m/s at sea level in 20 °C air: (a) neglecting radiation, (b) assuming an emissivity of 0.05.

10.4. A thermometer 4 mm in diameter reads 43.4 °C when held in a 61-m/s stream of air. What are the dynamic and stagnation temperatures of the air: (a) neglecting radiation, (b) assuming the thermometer radiates with an emissivity of 0.90 to surroundings at 65.6 °C?

10.5. At what velocity will aerodynamic heating become important (5% effect) in a tubular heat exchanger with air at 40 °C and 1 atmosphere and with a wall temperature of 65 °C?

10.6. A shock wave is passed through air at 12 °C and atmospheric pressure in a long tube. The velocity of the wave is measured and found to be 1.83 km/s. What maximum temperature might a tiny thermocouple attain in the gas behind the wave? Assume that the thermocouple is sufficiently small so that radiation and the effect of the thermocouple on the wave may be neglected.

10.7. A reciprocating pump at ground level is pumping 21 °C water from a reservoir with a surface level 20 feet below the ground through a 52.5-mm pipe with an equivalent length of 12.2 m. The net positive suction head of the pump is 0.6 m. What is the maximum rate of pumping?

10.8. Air at 40 °C is flowing over a cylinder 50 mm in diameter at a velocity of 12 m/s. Calculate the stagnation pressure and temperature and the actual temperature and pressure at the forward point of the cylinder.

10.9. Air in a reservoir at 290 K and 1 MPa is accelerated to $M_1 = 2$ in a nozzle. What are the pressure, temperature, and density?

10.10. Expand Equation (10.18) in a power series in M_1.

10.11. A centrifugal pump produces a head of 6.1 m when pumping water. Calculate the developed pressure. Estimate the developed head and pressure when pumping an oil with a density of 0.8 Mg/m^3.

10.12. The stagnation pressure obtained by setting $u = 0$ and $z = z_1$ in Equation (10.1) is

$$P_S = P_1 + \frac{\rho u_1^2}{2}$$

Compare with Equation (10.16) and explain.

10.13. A pitot tube reads the difference between the stagnation and the flowing pressure heads. If the reading is 100 mm of mercury in a stream of air at 21 °C calculate the velocity. Calculate the velocity using the equation in Problem 10.12 and compare.

10.14. A pitot tube is inserted in a stream of air with a velocity of 460 m/s and a temperature of 5 °C. A standing shock wave may be presumed to form ahead of the tube. What will be the reading on the pitot tube in millimeters of mercury? What are the stagnation and adiabatic-wall temperatures?

10.15. What is the maximum allowable speed of rotation of a 1.8 m propeller 6 m below the surface of 15 °C water?

10.16. A 152-mm pipeline carrying gasoline with a density of 0.849 Mg/m^3 and a vapor pressure of 6.9 kPa passes over a hill before terminating in a storage tank. The top of the hill is 12.2 m above the liquid level in the tank and the equivalent length of pipe between the hill and the tank is 1.6 km. Will the gasoline vaporize in the pipe line? Support your answer quantitatively.

10.17. Calculate the stagnation temperature, pressure and density for air flowing out of a 1.6-km-long, 52.5-m pipe at M = 0.8 and −1.6 °C.

10.18. Does the stagnation-temperature change in adiabatic flow if c_p changes? If a chemical reaction occurs? If work is done? Explain.

10.19. Does the invariance of the stagnation pressure in adiabatic reversible flow require constant c_p, ideal gas behavior and no work? Explain.

10.20. Can cavitation occur in a pipeline of constant diameter? Explain briefly.

10.21. Can cavitation occur in a reciprocating pump? Explain briefly.

10.22. Air is expanding from a vessel at 1 MPa to the atmosphere through a converging-diverging nozzle. The temperature, measured in the throat by a thermometer, is 38 °C.

a) What is the temperature of the air at the throat assuming that radiative transfer is negligible?

b) What is the reservoir temperature?

c) What is the pressure at the throat?

10.23. Air is expanding adiabatically from a vessel at 289 K and 1 MPa to the atmosphere through a converging-diverging nozzle.

a) Calculate the temperature and pressure at the throat.

b) Calculate the temperature read by a thermometer at the throat of the nozzle, neglecting radiative transfer.

10.24. Calculate the stagnation and adiabatic wall temperatures for 25 °C air with a velocity of 240 m/s. A turbulent boundary layer and Pr = 0.71 may be assumed.

References

1. R. T. Knapp and A. Hollander, "Laboratory Investigations of the Mechanism of Cavitation," *Trans. ASME* **70**, 419-435 (1948).
2. E. R. G. Eckert and R. M. Drake, Jr., *Heat and Mass Transfer*, McGraw-Hill, New York (1959), p. 265f.
3. R. W. Truitt, *Fundamentals of Aerodynamic Heating*, Ronald Press, New York (1960).

Nomenclature

A	area, m^2
a	acoustic velocity, m/s
B	constant of integration in Equation (1.36)
b	width of slot, m; coefficient in Equation (4.21), m/s · Pa^n or $m^{n+1} s^{2n-1}/kg^n$
C	discharge coefficient
$\mathbf{C_i}$	concentration of ith species, mol/m^3
c	specific heat capacity, J/kg · K or m^2/s^2 · K
D	diameter, m
d	height or depth, m
E	bulk modulus, kg/m · s^2
e	roughness of pipe, m
F	coefficient in Equations (8.26) – (8.29)
F_D	drag force of atmosphere, N or m · kg/s^2
Fr	Froude number = u^2/gd
f	friction factor = $\tau/\rho u^2$
G	mass velocity = ρu_b, kg/m^2 · s
g	acceleration (or specific force) due to gravity, m/s^2 or N/kg
H	specific enthalpy, J/kg or m^2/s^2
h	heat transfer coefficient = $j/(T_w - T_{aw})$, W/m^2 · K or kg/s^3 · K or J/m^2 · s · K
I	specific impulse, N · s/kg or m/s
j	heat flux density, W/m^2 or kg/s^3 or J/m^2 · s
K_{jk}	loss coefficient for source j based on reference velocity $u_k = 2lw_j/u_k^2$
$\mathbf{K_j}$	equilibrium constant for j^{th} reaction based on concentrations
k	exponent in Equation (1.46)
k	second-order reaction rate constant; m^3/mol · s
L	length of pipe or channel, m
L	equivalent length of pipe, m
ln	natural logarithm of
lw	lost work or viscous loss per mass of fluid, J/kg or m^2/s^2 or Nm/kg
M	Mach number = u/a
m	molar mass, kg/mol

NPSH	net positive suction head of pump, m
n	exponent of pressure for propellant gasification in Equation (4.21)
P	absolute pressure Pa or N/m^2 or kg/kg/m · s
P	vapor pressure, Pa
Pr	Prandtl number = ν/a
p_w	wetted perimeter of channel, m
Q_l	lower heating value of fuel, J/kg or m^2/s^2
q	heat entering system per unit mass, J/kg or m^2/s^2
R	universal gas law constant = 8.3143 J/mol · K or kg · m^2/mol · s^2 · K or m · N/mol · K or Pa · m^3/mol · K
r	reaction rate in Equation (7.37), mol/m^3 · s
$\bar{r}$	linear rate of gasification of solid propellant, m/s
S	specific entropy, J/kg · K or m^2/s^2 · K
s	width of channel, m
T	absolute temperature, K
T	period of a standing wave, s
t	time, s
U	specific internal energy = H − P/ρ, J/kg or m^2/s^2 or Pa · m^3/kg
u	velocity, m/s
u*	effective velocity, m/s
V	volume, m^3
v	volumetric flow rate, m^3/s
W	mass, kg
w	mass flow rate, kg/s
w	work done by system per unit mass, J/kg or m^2/s^2
x	coordinate in horizontal direction, m; length of pipe, m
y	distance from surface of liquid, m
Z	compressibility factor in Equation (3.9a)
z	coordinate in vertical direction, m

Greek letters

a	thermal diffusivity, m^2/s
β	heat capacity ratio function = $(\gamma+1)/(\gamma-1)$
γ	heat capacity ratio = c_p/c_V
Δ	finite difference between final and initial states or exit and entering states
ΔH_R	specific enthalpy change due to reaction, J/kg or m^2/s^2
$-\Delta P_f$	pressure drop due to friction, Pa or kg/m · s^2
ϵ	emissivity of surface
η	efficiency
θ	total exit angle of a diverging nozzle; angle of inclination of a channel

	from the horizontal, rad
λ	wavelength, m
μ	(dynamic) viscosity, Pa · s or kg/m · s
ν	kinematic viscosity = μ/ρ, m^2/s
ρ	density, kg/m^3
σ	surface tension (for waves), N/m or kg/s^2; Stefan coefficient (for radiation) = 5.6698×10^{-8} $W/m^2 \cdot K^4$ or $kg/s^3 \cdot K^4$ or $J/s \cdot m^2 \cdot K^4$
T	thrust, N or $kg \cdot m/s^2$
τ	shear stress imposed on wall by fluid or vice versa, Pa or $kg/m \cdot s^2$ or N/m^2
ϕ	mass ratio of fuel to air = w_f/w_a
ψ	angle of rocket motion relative to vertical, rad

Subscripts

a	air
atm	atmospheric
aw	adiabatic surface temperature
B	for Borda mouthpiece
b	bulk or mixed-mean
C	for sudden contraction
c	critical value for acoustic velocity; centerline
d	gate
E	for sudden expansion
e	at exit on velocity; equivalent on D
F	final (burnout); froth (on sieve tray)
f	fuel (on w); reacting (on A), friction (on P)
g	gas phase at inlet; i^{th} species with concentration
J	propulsion
j	source of lost work; j^{th} equilibrium constant
k	reference velocity for loss coefficient
L	liquid
lm	logarithmic mean = $(x_2 - x_1)/\ln x_2/x_1)$
m	mean
max	maximum
min	minimum
n	nozzle
O	overall
o	orifice
P	constant pressure
p	propellant
R	rocket; reacted mixture

r	of reflected wave
rev	reversible
S	constant entropy
s	stagnation or reservoir (u = 0)
T	constant temperature
t	at throat
U	unreacted mixture
V	constant volume (density)
v	Venturi
w	wave (on velocity); wall (on temperature); wetted (on perimeter)
z	over weir
ρ	constant density
0	initial or inlet or reference state
1,2,3...	designated states or locations
∞	ambient or external state

Superscripts

+	velocity relative to surroundings (for a rocket)
′	velocity relative to wave
″	velocity relative to reflected wave

The International System of Units (SI)

Base Units

Quantity	**Name**	**Symbol**
length	meter	m
mass	kilogram	kg
time	second	s
electric current	ampere	A
temperature	kelvin	K
amount of substance*	mole	mol
luminous intensity	candela	cd

**The mole is based on the gram instead of the kilogram.*

Derived Units with Special Names

Quantity	**Name**	**Symbol**	**Other Units**	**Base Units**
frequency	hertz	Hz		s^{-1}
force	newton	N		$m \cdot kg/s^2$
pressure, stress	pascal	Pa	N/m^2	$kg/m \cdot s^2$
energy, work, heat	joule	J	$N \cdot m$	$m^2 \cdot kg/s^2$
electric potential	volt	V	W/A	$m^2 \cdot kg/s^3 A$
electric charge	coulomb	C		$A \cdot s$
power, radiant flux	watt	W	J/s	$m^2 \cdot kg/s^3$

Derived Units Using Special Names

Quantity	Symbol	Base Units
dynamic viscosity	Pa · s	kg/m · s
surface tension	N/m	kg/s^2
heat flux density, irradiance	W/m^2	kg/s^3
heat capacity, entropy	J/K	$m^2 kg/s^2 \cdot K$
specific heat capacity, specific entropy	J/K · kg	$m^2/s^2 \cdot K$
specific energy	J/K	m^2/s^2
thermal conductivity	W/m · K	$m \cdot kg/s^3 \cdot K$
energy density	J/m^3	$kg/m \cdot s^2$
molar energy	J/mol	$m^2/kg \cdot s^2 \cdot mol$
molar entropy, heat capacity	J/mol · K	$m^2/kg \cdot s^2 \cdot mol \cdot K$

Supplementary Units

Quantity	Name	Symbol
plane angle	radian	rad
solid angle	steradian	sr

Derived Units from Supplementary Units

Quantity	Using Special Names	Base Units
angular velocity	–	rad/s
angular acceleration	–	rad/s^2
radiant intensity	W/sr	$m^2 \cdot kg/sr \cdot s^3$
radiance	$W/m^2 \cdot sr$	$kg/sr \cdot s^3$

Multiples and Submultiples

Factor	Prefix	Symbol
10^{12}	tera	T
10^{9}	giga	G
10^{6}	mega	M
10^{3}	kilo	k
10^{2}	hecto	h
10	deca	da
10^{-1}	deci	d
10^{-2}	centi	c
10^{-3}	milli	m
10^{-6}	micro	μ
10^{-9}	nano	n
10^{-12}	pico	p
10^{-15}	femto	f
10^{-18}	atto	a

Units Which May Be Used With the SI Units

Name	Symbol	Definition
minute	min	60s
hour	h	3600s
day	d	86400s
degree (of angle)	°	$(\pi/4)$rad
minute (of angle)	′	$(\pi/10800)$ rad
second (of angle)	″	$(\pi/648000)$ rad
atmosphere (standard)	atm	101 325 Pa
temperature–Celsius	°C	$T_{°C} = T_K - 273.15$

Constants in SI Units

k	Boltzmann constant	$1.380\,54 \times 10^{-23}$	J/K
h	Planck constant	$6.625\,6 \times 10^{-34}$	J · A
c	speed of light	$2.997\,925 \times 10^{8}$	m/s
N_A	Avagadro constant	$6.022\,52 \times 10^{23}$	mol^{-1}
V_m	molar volume at standard temperature and pressure	$2.241\,36 \times 10^{-2}$	m^3/mol

σ	Stefan constant	5.669 8	x 10^{-8}	$W/m^2 \cdot K^4$
R	molar gas constant	8.314 3		$J/mol \cdot K$
g	acceleration due to gravity (standard)	9.80665		m/s^2
G	gravitational constant	6.670	x 10^{-11}	$N \cdot m^2/kg^2$
$\lambda_{max}T$	$\lambda_{max}T$ (Wein's Law)	2.8978	x 10^{-3}	$m \cdot K$
	specific heat capacity of water at 15 °C	4.1855	x 10^3	$J/kg \cdot K$
	velocity of sound in air at standard temperature and pressure	333.46		m/s
	viscosity of water at 20 °C	1.0019	x 10^{-3}	$Pa \cdot s$
	viscosity of air at 20 °C	1.81	x 10^{-5}	$Pa \cdot s$
	charge on an electron	−1.602 10	x 10^{-19}	C
	standard atmosphere	1.013 25	x 10^5	Pa
	standard temperature	273.15		K

Conversion Factors

Quantity	**Symbol or Units**	**Definition**
inch	in	2.54 x 10^{-3} m
foot	ft	0.3048 m
yard	yd	0.9144 m
mile	mi	1609.344 m
circular mil	mil	5.067 025 x 10^{-10} m
micron	μ	10^{-6} m
ångström	Å	10^{-10} m
cubic foot	ft^3	2.831 865 x 10^{-2} m^3
gallon	gal	3.785 412 x 10^{-3} m^3
barrel (42 gallon)	bbl	0.158 987 3 m^3
liter	l	1 dm^3 or 10^{-3} m^3
mile/hour	mi/hr	0.447 4 m/s
gallon/minute	gpm	6.309 020 x 10^{-5} m^3
pound	lb	0.453 592 37 kg
temperature	°R	$T_{°R}/1.8 = T_K$
temperature	°C	$T_{°C} + 273.15 = T_K$
temperature	°F	$(T_{°F} + 459.67)/1.8 = T_K$
British thermal unit	BTU	1055.056 J
calorie	cal	4.1868 J
erg	erg	10^{-7} J
dyne	dyn	10^{-5} N

kilogram-force	kgf	9.806 65 N
pound-force	lbf	4.448 22 N
pressure (pound-force/inch^2)	psi	6.894 757 x 10^3 Pa
manometric pressure	mmHg	133.322 Pa
standard atmosphere	atm	1.013 25 x 10^5 Pa
head of water	ft	2.988 98 x 10^2 Pa
horsepower	HP	7.456 999 x 10^2 W
viscosity	lb/ft · s	1.488 164 Pa · s
centipoise	cp	1.000 000 x 10^{-3}
centistoke	cs	1.000 000 x 10^{-6} m^2/s
heat flux density	BTU/hr · ft^2	3.154 603 W/m^2
thermal conductivity	BTU/hr · ft · °R	1.730 734 W/m · K
heat transfer coefficient	BTU/hr · ft^2 · °R	5.678 263 W/m^2 · K
heat capacity	BTU/lb · °F, cal/gm · °C	4.1868 x 10^3 J/kg · K
thermal diffusivity	ft^2/hr	2.580 64 m^2/s

Index of Authors*

(asterick indicates complete reference)

Index of Subjects

gravitational waves (of). 113f
long waves (of) 118
maximum
in a nozzle 12,23f,35
in open channel flow 113f
in a pipe 56f
to avoid cavitation. 122f
to avoid vapor-binding 127
minimum for hydraulic jump. . . 116
minimum for significant aero-dynamic heating 130f
particle 23
detonation waves (for) . 92,108-110
reflected waves (for) 85

Stuart Winston Churchill

was born in Imlay City, Michigan. He attended the University of Michigan where he received bachelor's degrees in both mathematics and chemical engineering. After 5 years with Shell Oil and Frontier Chemical he returned to the University of Michigan for graduate work, continuing as a faculty member and eventually becoming Chairman of Chemical and Metallurgical Engineering. In 1967 he accepted the Carl V.S. Patterson Professorship at the University of Pennsylvania, which has permitted him to devote his efforts to teaching, research and writing on the rate processes. In 1976, the students in all of engineering selected him for the Reid S. Warren, Jr., Award for excellence in teaching. He served as the President of the AIChE in 1966 and has received their Professional Progress, Wilhelm H. Walker and Warren K. Lewis Awards. He is also a member of the American Chemical Society, the Combustion Institute and the National Academy of Engineering.